VOLUME SEVEN

DEVELOPMENTS IN NEUROETHICS AND BIOETHICS

Brains and Machines:
Towards a Unified Ethics of
AI and Neuroscience

Serial Editor

Judy Illes, CM, PHD
UBC Distinguished University Scholar
UBC Distinguished Scholar in Neuroethics
Division of Neurology, Department of Medicine
Neuroethics Canada
The University of British Columbia
2211 Wesbrook Mall, Koerner S124
Vancouver BC V6T 2B5
Canada

Katherine Bassil, PHD, Assistant to the Editor
Department of Psychiatry and Neuropsychology
Maastricht University
Maastricht, the Netherlands
Department of Bioethics and Health Humanities
Julius Center, University Medical Center Utrecht
Utrecht, The Netherlands

VOLUME SEVEN

Developments in
NEUROETHICS AND BIOETHICS

Brains and Machines: Towards a Unified Ethics of AI and Neuroscience

Edited by

MARCELLO IENCA

*Institute for History and Ethics of Medicine,
School of Medicine and Health,
Technical University of Munich, Germany;
College of Humanities, EPFL, Switzerland*

GEORG STARKE

*Institute for History and Ethics of Medicine,
School of Medicine and Health,
Technical University of Munich, Germany;
College of Humanities, EPFL, Switzerland*

ACADEMIC PRESS
An imprint of Elsevier

Academic Press is an imprint of Elsevier
125 London Wall, London, EC2Y 5AS, United Kingdom
50 Hampshire Street, 5th Floor, Cambridge, MA 02139, United States
525 B Street, Suite 1650, San Diego, CA 92101, United States

First edition 2024

Copyright © 2024 Elsevier Inc. All rights are reserved, including those for text and data mining, AI training, and similar technologies.

Publisher's note: Elsevier takes a neutral position with respect to territorial disputes or jurisdictional claims in its published content, including in maps and institutional affiliations.

No part of this publication may be reproduced or transmitted in any form or by any means, electronic or mechanical, including photocopying, recording, or any information storage and retrieval system, without permission in writing from the publisher. Details on how to seek permission, further information about the Publisher's permissions policies and our arrangements with organizations such as the Copyright Clearance Center and the Copyright Licensing Agency, can be found at our website: www.elsevier.com/permissions.

This book and the individual contributions contained in it are protected under copyright by the Publisher (other than as may be noted herein).

Notices
Knowledge and best practice in this field are constantly changing. As new research and experience broaden our understanding, changes in research methods, professional practices, or medical treatment may become necessary.

Practitioners and researchers must always rely on their own experience and knowledge in evaluating and using any information, methods, compounds, or experiments described herein. In using such information or methods they should be mindful of their own safety and the safety of others, including parties for whom they have a professional responsibility.

To the fullest extent of the law, neither the Publisher nor the authors, contributors, or editors, assume any liability for any injury and/or damage to persons or property as a matter of products liability, negligence or otherwise, or from any use or operation of any methods, products, instructions, or ideas contained in the material herein.

ISBN: 978-0-443-15869-8
ISSN: 2589-2959

For information on all Academic Press publications
visit our website at https://www.elsevier.com/books-and-journals

Publisher: Zoe Kruze
Acquisitions Editor: Mariana Kuhl
Editorial Project Manager: Akanksha Marwah
Production Project Manager: Maria Shalini
Cover Designer: Vicky Pearson

Typeset by MPS Limited, India

Contents

Contributors	*xi*
About the Editors	*xv*
Introduction: Navigating ethics at the intersection of AI and neuroscience	*xix*

Section 1 AI for neuroscience

1. AI for brain-computer interfaces   3

David Haslacher, Tugba Basaran Akmazoglu,
Amanda van Beinum, Georg Starke, Maria Buthut, and
Surjo R. Soekadar

1.	Introduction	4
2.	Basic types and applications of brain-computer interfaces	6
3.	AI-enhanced brain-computer interfaces	9
4.	Scope and scalability of AI-enhanced BCIs	14
5.	Perspectives and ethical implications	16
6.	Conclusions	19
	References	20

2. Computational psychiatry and AI - High hopes: heralded heights or hollow hype?   29

Derya Şahin

1.	Introduction	29
2.	General AI-related ethical concerns	31
3.	Ethical concerns specific to computational psychiatry	32
4.	Conclusion	42
	References	42

3. Computational psychiatry and digital phenotyping: Ethical and neurorights implications   47

José M. Muñoz, Diego Borbón, and Ana Maria Bezerra

1.	Introduction	48
2.	Computational psychiatry and its ethical challenges	48
3.	The case of digital phenotyping: From health applications to the consumer domain	51

4. Potential implications for neurorights	54
5. Final remarks: Towards the integration of different data levels	57
References	58

Section 2 Neuroscience for AI

4. Neuroscience for AI: The importance of theory of mind — 65

Christelle Langley, Fabio Cuzzolin, and Barbara J. Sahakian

1. Introduction	66
2. *Hot* and *cold* cognition	67
3. *Hot* and *cold* cognition in theory of mind	71
4. *Hot* and *cold* cognition and theory of mind in AI	73
5. Limitations and ethical concerns to theory of mind in AI	76
6. Conclusions	78
References	79

5. Sense of agency in human-human and human-computer interactions — 85

Sofia Bonicalzi

1. Introduction	86
2. The sense of agency within the cognitive (neuro)science of action	86
3. Interacting with humans affects the SoA	90
4. Interacting with artificial devices and AI affects the SoA	92
5. Conclusions	97
References	98

6. Anthropomorphism in social AIs: Some challenges — 101

Arleen Salles and Abel Wajnerman Paz

1. Introduction	101
2. Anthropomorphism	103
3. Social chatbots	104
4. Two strategies	106
5. Pragmatic strategy: Some objections	108
6. Conclusion	114
References	114

Contents   vii

7. (Mis)decoding affect in the face and in the brain   119
Marco Viola

1. QWERTY keyboards and affective science	119
2. Decoding emotion from faces	125
3. Decoding emotion from (neurovascular proxies for) brain data	129
4. Summary and open problems	134
References	135

Section 3 Finding common ground

8. Algorithmic regulation: A compatible framework for AI and DTC neurotechnologies   143
Lucille Nalbach Tournas and Walter G. Johnson

1. Introduction	144
2. Direct to consumer neurotechnologies	145
3. Algorithmic regulation	147
4. Applying algorithmic regulation to neurotechnology	149
5. Conclusion	156
References	157
Further reading	159

9. The extended mind thesis and the cognitive artifacts approach: A comparison   161
Guido Cassinadri and Marco Fasoli

1. Introduction	162
2. The extended cognition and the cognitive artifacts approach	163
3. How to extend cognition, cognitive artifacts, and AI	167
4. Conclusion: Compatibility and differences between EXT and CA and some ethical implications	177
References	179

10. The ethical implications of indicators of consciousness in artificial systems   185
Michele Farisco

1. Introduction	185
2. The prism of consciousness	187
3. Indicators of consciousness and their application to AI	190

4. Ethical implications of indicators of consciousness in AI	193
5. Conclusion	196
References	196

11. Moral dimensions of synthetic biological intelligence: Unravelling the ethics of neural integration — 199

Masanori Kataoka, Christopher Gyngell, Julian Savulescu, and Tsutomu Sawai

1. Introduction	200
2. Moral standing-related ethical dimensions of SBI systems	202
3. Non moral standing-related ethical dimensions of SBI systems	207
4. Conclusion	210
References	211

12. What the embedded ethics approach brings to AI-enhanced neuroscience — 215

Stuart McLennan, Theresa Willem, and Amelia Fiske

1. Introduction	215
2. Embedded ethics	218
3. Applying embedded ethics to AI-enhanced neuroscience	220
4. Conclusion	222
References	223

13. From being embedded in practice: Working situated and interdisciplinary in the neurosciences and neurocomputation as ethicists and social scientists — 225

Franziska B. Schönweitz, Anja K. Ruess, and Ruth Müller

1. Introduction	226
2. Situated knowledges and interdisciplinary research	228
3. Embedded ethics and social sciences in practice—active integration in frontier neuroscience and neurocomputation	230
4. Investigating neuroethics and ethics of AI within NEUROTECH: Data perception, security, and representativeness	232
5. The need for situatedness of ethics and social science research in complex network structures	236
6. Conclusion	238
References	238

Contents

ix

14. Beyond participation: Towards a community-led approach to value alignment of AI in medicine   **241**

Philipp Kellmeyer

1. Introduction	242
2. Conceptualizing participation and partaking	243
3. Methods of participatory research	247
4. Current paradigms for value alignment: Ethics-by-design and embedded ethics	251
5. Towards community-led value alignment in medical AI development	252
6. Limits of participatory methods in medical research and AI development	256
7. Conclusions	258
References	259

Epilogue: Harmonizing the ethical symbiosis of brains and machines   *263*

Contributors

Tugba Basaran Akmazoglu
Faculty of Law, University of Ottawa, Ottawa, ON, Canada

Ana Maria Bezerra
Unichristus School of Law, Unichristus, Fortaleza, CE, Brazil

Sofia Bonicalzi
Department of Philosophy, Communication and Performing Arts, Roma Tre University, Rome, Italy; CVBE Cognition, Value and Behavior, Ludwig-Maximilians-Universität München, Munich, Germany

Diego Borbón
Center for Studies on Genetics and Law, Universidad Externado de Colombia, Bogotá, Colombia

Maria Buthut
Clinical Neurotechnology Laboratory, Department of Psychiatry and Neurosciences, Charité Campus Mitte (CCM), Charité—Universitätsmedizin Berlin, Germany

Guido Cassinadri
Institue of Health Science, Sant'Anna School of Advanced Studies, Pisa, Italy

Fabio Cuzzolin
School of Engineering, Computing and Mathematics, Oxford Brookes University, Oxford, United Kingdom

Michele Farisco
Centre for Research Ethics and Bioethics, Uppsala University, Sweden; Biogem, Biology and Molecular Genetics Research Institute, Ariano Irpino, Italy

Marco Fasoli
Department of Philosophy, University of Roma La Sapienza, Rome, Italy

Amelia Fiske
Munich Embedded Ethics Lab (MEEL), Institute of History and Ethics in Medicine, Department of Clinical Medicine, TUM School of Medicine and Health, Technical University of Munich, Munich, Germany

Christopher Gyngell
Murdoch Children's Research Institute; Department of Paediatrics, University of Melbourne, Melbourne, Australia

David Haslacher
Clinical Neurotechnology Laboratory, Department of Psychiatry and Neurosciences, Charité Campus Mitte (CCM), Charité—Universitätsmedizin Berlin, Germany

Walter G. Johnson
School of Regulation and Global Governance (RegNet), The Australian National University, Canberra ACT, Australia

Masanori Kataoka
Graduate School of Humanities and Social Sciences, Hiroshima University, Higashi-Hiroshima, Japan

Philipp Kellmeyer
Data and Web Science Group, School of Business Informatics and Mathematics, University of Mannheim, Mannheim, Germany; Human-Technology Interaction Lab, University of Freiburg-Medical Center, Freiburg im Breisgau, Germany; Institute for Biomedical Ethics and History of Medicine, University of Zurich, Zurich, Switzerland

Christelle Langley
Department of Psychiatry, University of Cambridge, Cambridge, United Kingdom

Stuart McLennan
Munich Embedded Ethics Lab (MEEL), Institute of History and Ethics in Medicine, Department of Clinical Medicine, TUM School of Medicine and Health, Technical University of Munich, Munich, Germany

José M. Muñoz
Kavli Center for Ethics, Science, and the Public, University of California, Berkeley, CA, United States

Ruth Müller
Department of Science, Technology and Society (STS), TUM School of Social Sciences and Technology; Department of Economics and Policy, TUM School of Management, Technical University of Munich, Munich, Germany

Anja K. Ruess
Department of Science, Technology and Society (STS), TUM School of Social Sciences and Technology; Department of Economics and Policy, TUM School of Management, Technical University of Munich, Munich, Germany

Barbara J. Sahakian
Department of Psychiatry, University of Cambridge, Cambridge, United Kingdom

Derya Şahin
Stiftung Mercator, Essen, Germany

Arleen Salles
Institute of Neuroethics (IoNx), Think and Do Tank, Atlanta, Georgia, USA

Julian Savulescu
Murdoch Children's Research Institute, Melbourne, Australia; Faculty of Philosophy, University of Oxford, Oxford, Uinted Kingdom; Centre for Biomedical Ethics, Yong Loo Lin School of Medicine, National University of Singapore, Singapore, Singapore

Tsutomu Sawai
Graduate School of Humanities and Social Sciences, Hiroshima University, Higashi-Hiroshima, Japan; Centre for Biomedical Ethics, Yong Loo Lin School of Medicine, National University of Singapore, Singapore, Singapore; Institute for the Advanced Study of Human Biology (ASHBi), Kyoto University, Kyoto, Japan

Contributors

Franziska B. Schönweitz
Institute of History and Ethics in Medicine, TUM School of Medicine and Health, TUM School of Social Sciences and Technology, Technical University of Munich, Munich, Germany

Surjo R. Soekadar
Clinical Neurotechnology Laboratory, Department of Psychiatry and Neurosciences, Charité Campus Mitte (CCM), Charité—Universitätsmedizin Berlin, Germany

Georg Starke
Institute for History and Ethics of Medicine, School of Medicine and Health, Technical University of Munich, Germany; Collège des Humanités, École Polytechnique Fédérale de Lausanne, Switzerland

Lucille Nalbach Tournas
School of Life Sciences, Arizona State University, Tempe, AZ, United States

Amanda van Beinum
Faculty of Law, University of Ottawa, Ottawa, ON, Canada

Marco Viola
Department of Philosophy, Communication, and Performing Arts, University of Roma Tre, Rome, Italy

Abel Wajnerman Paz
Pontificial Universidad Catolica de Chile, Instituto de Eticas Aplicadas, Santiago de Chile, Chile

Theresa Willem
Munich Embedded Ethics Lab (MEEL), Institute of History and Ethics in Medicine, Department of Clinical Medicine, TUM School of Medicine and Health, Technical University of Munich, Munich, Germany

About the Editors

Marcello Ienca is a Professor of Ethics of Artificial Intelligence and Neuroscience, and the Deputy director of the Institute of Ethics and History of Medicine at the Technical University of Munich (TUM), Germany. He is also heading the Intelligent Systems Ethics unit at the EPFL, Switzerland. Ienca currently serves as the Lead of the Neuroethics Working Group of the International Brain Initiative and is a member of the Board of Directors of the International Neuroethics Society. He represents Switzerland at the OECD in the production of the Guidelines on Responsible Innovation in Neurotechnology. Dr. Ienca is one of the 24 international experts appointed by UNESCO's Director-General to the Ad Hoc Expert Group (AHEG), tasked with drafting the upcoming Recommendation on the Ethics of Neurotechnology. Additionally, he has served as an expert to the Advisory Committee of the UN Human Rights Council during the development of the report related to the motion resolution on neurotechnologies and has authored multiple reports on Neurotechnology and Human Rights for the Council of Europe. Dr. Ienca's research has been featured in academic journals such as Neuron, Nature Biotechnology, Nature Machine Intelligence, and Nature Medicine, as well as media outlets such as Nature, The New Yorker, The Guardian, The Times, Die Welt, The Independent, and the Financial Times. He is the co-editor of The Routledge Handbook of the Ethics of Human Enhancement (Routledge, 2024), The Cambridge Handbook of Information Technology, Life Sciences, and Human Rights (Cambridge University Press, 2022), and Intelligent Assistive Technologies for Dementia (Oxford University Press, 2019).

Dr. med. Dr. phil. Georg Starke, MPhil
Institute for History and Ethics of Medicine, Technical University of Munich, Germany, and College of Humanities, EPFL, Switzerland.

Georg Starke is a Research Associate at the Chair for Ethics of AI and Neuroscience at the Institute for History and Ethics of Medicine at the Technical University of Munich and Postdoctoral Researcher in the Intelligent Systems Ethics Group at the Swiss Federal Institute of Technology in Lausanne (EPFL). Trained as philosopher and physician, he conducts research at the intersection of AI ethics, medical ethics, and neuroscience, relating philosophical and ethical questions to empirical work in the field.

Dr. Judy Illes, CM, PhD, FCAHS, FRSC, Division of Neurology, Department of Medicine, University of British Columbia, Vancouver, BC, Canada.

Dr. Illes is Professor of Neurology at the University of British Columbia (UBC), UBC Distinguished University Scholar, Distinguished University Scholar in Neuroethics, and Director of Neuroethics Canada. She is a pioneer of the field of neuroethics through which she has made groundbreaking contributions to crosscultural ethical, legal, social, and policy challenges at the intersection of the brain sciences and biomedical ethics. She was awarded the Order of Canada, one of the country's highest recognition of its citizens, in 2017.

Katherine Bassil, PhD Department of Psychiatry and Neuropsychology, Maastricht University, Maastricht, The Netherlands; Department of Bioethics and Health Humanities, Julius Center, University Medical Center Utrecht, Utrecht, The Netherlands.

Katherine Bassil is currently an Assistant Professor in neuroethics at the Julius Center, University Medical Center Utrecht (UMCU) in the Netherlands. She holds a second affiliation at Maastricht University, at the Department of Psychiatry and Neuropsychology. Katherine is involved in several international organizations including the IEEE Neuroethics Framework and the International Neuroethics Society (INS). She is also the founder of Neuroethics Today, an educative platform raising awareness on neuroethics topics.

Introduction: Navigating ethics at the intersection of AI and neuroscience

Georg Starke[a,b]

[a]Institute for History and Ethics of Medicine, Technical University of Munich, Germany
[b]College of Humanities, EPFL, Switzerland

Every discipline tells a story: where it comes from, what it is and where it is going.
Simon Schaffer, How Disciplines Look

The history and development of two key technologies of the 21st century, artificial intelligence (AI) and neuroscience, are deeply intertwined. Many of the initial pioneers bringing about today's AI systems had substantive expertise in neuroscience and psychology, and collaborative efforts between the disciplines yielded significant scientific progress (Hassabis et al., 2017). Some of the terminology that computer science inherited from neuroscience and psychology provides testimony to this development, from (artificial) neural networks to the term intelligence itself. In turn, AI increasingly comes to the aid of neuroscience, providing tools for the advancement of neuroscience (Macpherson et al., 2021). Not only can AI help to model cortical processes more accurately than ever before but it can also support the interpretation of neural data to a degree that seemed out of reach just a decade ago. Latent diffusion models now allow a high-resolution reconstruction of images from functional magnetic resonance imaging (Takagi & Nishimoto, 2023) and specific songs such as Pink Floyd's *Another Brick in the Wall (Part 1)* can be reconstructed from human auditory cortex activity (Bellier et al., 2023).

Given these and many other well-known advancements, it comes as no surprise that there is burgeoning public and scholarly interest both in AI and in AI-enabled neuroscience, and what their progress may mean for individuals and societies interacting with them. Recent examples include debates about supposedly conscious large language models (Ienca, 2023), sentient brain organoids (Bassil, 2023; Milford et al., 2023) or AI-enabled neurotechnology preparing to leave the clinical domain (Schnabel, 2023). Against this background, a joined ethical framework for reflecting on neuroscience and AI seems more crucial than ever. Yet, despite the increasing intersection between neuroscience and AI and the fruitful interactions between the two scientific domains, ethical discourse about neuroscience and AI has more often than not moved in parallel. This may be little surprising, for as Simon Schaffer reminds us, each field and discipline provides narratives, tells its own story, and *disciplines* its members, training them in particular methods and concepts (Schaffer, 2013). Oddly but evidently, this seems to be the case even for highly interdisciplinary fields themselves (Schaffer, 2013), with AI ethics and neuroethics providing no obvious exception.

When it comes to AI and neuroscience, there is a material need though to cross disciplinary boundaries and bring together complementary debates and rivalling viewpoints. Speaking from their respective disciplinary standpoints, warranted calls for learning from each other have already been put forth. Sara Berger and Francesca Rossi have suggested that neuroethics could learn from tech companies to move from abstract general principles to scaled industrial practice (Berger & Rossi, 2022). Prominent neuroethicists have argued in turn that AI ethics could learn important lessons from the historically older philosophical and ethical reflection on neuroscience, with view to methodology as well as shared questions pertaining to, for instance, consciousness, intelligence or the status of human beings (Farisco et al., 2022). We concur and believe that both are right. AI ethics and neuroethics need to join forces, bring together their expertise and learn from each other, in order to help shape the emerging technologies at the intersection of AI and neuroscience in a responsible manner. This volume aims to facilitate such exchanges between the two fields.

Brains and Machines: Towards a Unified Ethics of AI and Neuroscience therefore provides a comprehensive overview of concepts and ethical issues at the intersection of both domains. It assembles diverse voices from various disciplines to provide a more unified view onto joint ethical challenges, drawing on perspectives from two complementary lines of discourse. The volume is structured in three main sections. Section 1, *AI for Neuroscience* provides background by critically discussing the use of AI for neuroscientific applications, especially in clinical contexts from neurology and psychiatry. Section 2, *Neuroscience for AI* addresses the complementary perspective, discussing instances where findings and debates from neuroscience provide cues and inspiration to the study of AI. While both sections provide ethical discussions of the technologies in questions, they also offer definitions and clarify key conceptual issues that will help readers understand the philosophical, ethical, regulatory and social implications addressed in the subsequent chapters. Section 3, *Finding*

Introduction xxi

common ground, constitutes the final and broadest part of the volume. Bringing together scholars from various disciplines and backgrounds, this section aims towards shared conceptual frameworks, identifies ethical challenges at the intersection of neuroscience and AI, highlights challenges and opportunities of finding common ground in interdisciplinary settings, and proposes joint methodologies for addressing ethical questions in both AI and neuroscience.

The book's first section explores the utility and consequences of AI for (clinical) neuroscience. David Haslacher, Tugba Basaran Akmazoglu, Amanda van Beinum, Georg Starke, Maria Buthut and Surjo Soekadar initiate this exploration by delving into the integration of AI into clinical neurotechnology. Commencing with a historical overview of Brain-Computer Interface (BCI) development spanning from the 20th century to today's AI-enabled bidirectional BCIs, the chapter proceeds to introduce various BCI types and examines the clinical utility of AI-enhanced BCIs. This utility extends from restoring sensorimotor function and speech to treating affective disorders. After delineating the foreseeable scope and scalability of the technology, the chapter concludes by delving into ethical challenges specific to AI-enabled BCIs.

Expanding on this neurotechnological perspective, the subsequent two chapters delve into the utilisation of AI in psychiatry and its corresponding ethical and legal consequences. Both chapters specifically illuminate the challenges associated with employing AI in computational psychiatry, which integrates mathematical, statistical, and computational tools to advance knowledge and treatment of psychiatric disorders. Derya Şahin initiates this exploration by acquainting readers with the expanding field and mapping its ethical challenges. She distinguishes between ethical concerns general to AI in healthcare, such as privacy, fairness, and explainability, and challenges specific to AI in computational psychiatry, both theoretical and methodological. Deepening engagement with computational psychiatry, the third chapter, authored by José Manuel Muñoz, Diego Borbón, and Ana Maria Bezerra, investigates the potential of AI-enabled computational psychiatry for digital phenotyping and its ethical implications. As the authors illustrate, the discipline's pursuit of systematically analysing data obtained through wearable digital technology, such as smartphones, can serve clinical psychiatric purposes as well as non-clinical objectives like neuromarketing. The chapter offers an overview of the ensuing ethical and legal implications of these AI-enabled technologies, focusing particularly on a neurorights framework and addressing issues of privacy and algorithmic bias.

The volume's second section explores the role of neuroscientific knowledge for current AI development, shedding light on experience and knowledge that the historically older discipline can contribute to the improvement of AI models and to addressing the ethical challenges they pose. Christelle Langley, Fabio Cuzzolin and Barbara Sahakian initiate the section by emphasizing the significance of theory of mind. They underscore how a deeper understanding of this fundamental concept in cognitive science can enhance interactions between humans and future AIs. The authors present a thorough introduction to theory of mind, distinguishing between hot cognitive processes, which are emotionally charged, and cold cognitive processes, bare of emotional influence. They show how such findings from neuroscience could be useful for AI improvement and highlight ethical concerns arising from such neuroscience-inspired AI. A closely related question is also pursued in the following chapter. Drawing from debates within cognitive neuroscience and the philosophy of action, Sofia Bonicalzi elucidates what AI can glean from neuroscience regarding the sense of agency, i.e. the subjective feeling and judgement that one controls their own actions and, consequently, their outcomes. Bonicalzi provides an accessible introduction to this intricate empirical and conceptual debate, illustrating how incorporating a sense of agency into AI systems could reshape human interaction with these systems. The author reviews recent empirical evidence on the subject, culminating in an exploration of the implications for the ensuing sense of responsibility for actions in humans and, potentially, in machines.

The following chapter by Arleen Salles and Abel Wajnerman Paz takes a complementary view. Instead of using knowledge from neuroscience to teach machines how to better interact with humans, the chapter looks at our human perspective when dealing with AI. Using AI-enabled social chatbots as examples, the authors provide an insightful introduction to the function of and motivation for anthropomorphising non-human entities and delve into the motivations behind such practices. The chapter critically explores various justifications for anthropomorphism, differentiating between ontological and pragmatic strategies. Through this analysis, the authors advocate for a more nuanced consideration of the ontological, psychological, and ethical assumptions underpinning widespread anthropomorphisation of AI systems such as social chatbots. The section is rounded-off by Marco Violas chapter on affect recognition from neural activity and facial movements. The chapter traces challenges in contemporary AI technology back to a neuroscientific legacy, namely

Paul Ekman's taxonomy of six innate emotion categories. Despite being a contested theory likely requiring modification, the chapter illustrates how Ekman's ideas continue to shape current approaches to affect recognition, as enabled by modern AI models. It underscores the implications and risks associated with uncritically adopting such legacies, emphasizing the need for improved communication between the fields of AI and neuroscience to mitigate potential risks and ensure ethical progress.

The final and most extensive section of the book therefore endeavors to find common ground between AI ethics and neuroethics. The section's initial four chapters focus on conceptual questions and specific applications. Here, Lucy Tournas and Walter Johnson make a case for algorithmic regulation as a compatible framework for both AI and direct-to-consumer (DTC) neurotechnologies. Drawing on Karen Yeung's work, the authors compellingly demonstrate that both DTC neurotechnologies and AI influence, enable, and restrict preferences, behaviour, and values at individual and societal levels. By proposing a compatible framework, the authors provide a fresh perspective for analysing ethical, regulatory, political, and economic dimensions of these emerging technologies, with significant implications for key ethical concerns such as privacy, autonomy, and personal identity. Guido Cassinadri and Marco Fasoli subsequently illustrate how philosophical discussions about neurotechnology and AI share common ground. Their chapter introduces readers to two competing philosophical theories and their underlying argumentative strategies: the extended mind hypothesis, which posits that certain tools integral to cognitive processes should be considered part of the mind, and the cognitive artifact approach, which assigns such crucial tools to a specific class of objects. By comparing these theories and connecting them to AI, the authors reveal how a debate originating in the philosophy of mind remains relevant to reflections on both neurotechnological devices and AI. They outline how the choice of either theory as an analytical tool impacts the conceptualization of questions related to technology and argue that these perspectives should complement each other when addressing ethical issues associated with AI and neurotechnology. Staying with an equally fundamental philosophical question challenging neuroscience and AI, Michele Farisco tackles the topic of indicators of consciousness, their application to artificial systems, and the ensuing ethical implications. Beginning with a conceptual clarification of the meaning and characteristics of consciousness, Farisco puts forth a list of candidate indicators for consciousness applicable to artificial systems. The author critically reflects on the ethical implications of these indicators, with particular attention given to the conditions under which artificial systems might be considered moral subjects and emphasising the importance of non-anthropocentric approaches for a nuanced reflection on artificial consciousness. The urgency to engage in discussions surrounding indicators of consciousness and the moral status of artificial systems is underscored in the subsequent chapter authored by Masanori Kataoka, Christopher Gyngell, Julian Savulescu and Tsutomu Sawai. Turning to novel systems merging neuroscience and AI, the authors investigate the moral dimensions of Synthetic Biological Intelligence (SBI), which combines neural tissue with electrical arrays. The chapter provides a comprehensive exploration of the ethical challenges stemming from this hybrid technology. It distinguishes between issues related to the moral standing of SBI, such as consciousness, and ethical questions independent of it, such as consent from cell donors or the equitable sharing of benefits.

The final three chapters of the volume are concerned with finding methodological common ground shared by AI ethics and neuroethics, offering a forward-looking perspective by discussing suitable methodologies for developing ethically sound technologies. Two chapters delve into the influential embedded ethics approach. Stuart McLennan, Theresa Willem and Amelia Fiske emphasize the importance of not relegating ethics to an afterthought in the development of emerging technologies. They outline key elements of embedded ethics for addressing ethical, legal and social challenges and discuss the framework with particular view to AI-enabled neurotechnology. Moving from theory to practice, Franziska Schönweitz, Anja Ruess and Ruth Müller spell out what embedded ethics means in the field, drawing on their experiences as part of an embedded ethics research team at the intersection of neurotechnology and AI. They illuminate the challenges of interdisciplinary work, provide suggestions how to address them, and advocate for a situated perspective on the ethical and social challenges posed by emerging technologies. Complementary to these perspectives, Philipp Kellmeyer explores an alternative path to achieve value alignment in medical and neurotechnological AI systems. The chapter discusses established methodologies such as embedded ethics and proposes community-led approaches as an alternative. Following a systematic exploration of participation and partaking in research and development, the chapter delineates methods of participatory research in various areas. It compellingly argues for a model of community-led AI development in healthcare contexts, founded on four key pillars. After these hopeful outlooks to future value–aligned technological developments, the book concludes with an epilogue by Marcello Ienca, providing a reflective overview of the volume's chapters and topics.

In light of the rapidly evolving environments of neurotechnology and AI there is little doubt that ethical debates accompanying their development will remain crucial in the decades to come. We are confident that the contributions assembled here can provide a valuable contribution to public and scholarly debates surrounding these key technologies and hope that this volume will serve as both an entry point to the field and as asset for future research shaping the development of AI, neuroscience, and their intersection.

References

Bassil, K. (2023). The end of 'mini-brains'! Responsible communication of cerebral organoid research. *Molecular Psychology: Brain, Behavior, and Society, 2*, 13.

Bellier, L., Llorens, A., Marciano, D., Gunduz, A., Schalk, G., Brunner, P., & Knight, R. T. (2023). Music can be reconstructed from human auditory cortex activity using nonlinear decoding models. *PLoS biology, 21*(8), e3002176.

Berger, S. E., & Rossi, F. (2022). Addressing neuroethics issues in practice: Lessons learnt by tech companies in AI ethics. *Neuron, 110*(13), 2052–2056.

Farisco, M., Evers, K., & Salles, A. (2022). On the Contribution of neuroethics to the ethics and regulation of Artificial intelligence. *Neuroethics, 15*(1), 4.

Hassabis, D., Kumaran, D., Summerfield, C., & Botvinick, M. (2017). Neuroscience-inspired artificial intelligence. *Neuron, 95*(2), 245–258.

Ienca, M. (2023). Don't pause giant AI for the wrong reasons. *Nature Machine Intelligence,* 1–2.

Macpherson, T., Churchland, A., Sejnowski, T., DiCarlo, J., Kamitani, Y., Takahashi, H., & Hikida, T. (2021). Natural and Artificial Intelligence: A brief introduction to the interplay between AI and neuroscience research. *Neural Networks, 144*, 603–613.

Milford, S. R., Shaw, D., & Starke, G. (2023). Playing Brains: The Ethical Challenges Posed by Silicon Sentience and Hybrid Intelligence in DishBrain. *Science and Engineering Ethics, 29*(6), 38.

Schaffer, S. (2013). How disciplines look. In A. Barry, & G. Born (Eds.). *Interdisciplinarity: Reconfigurations of the Social and Natural Sciences* (pp. 57–81). Routledge.

Schnabel, U. (2023). Anschluss gesucht. *ZEIT,* 29–30.

Takagi, Y., & Nishimoto, S. (2023). High-resolution image reconstruction with latent diffusion models from human brain activity. Proceedings of the IEEE/CVF Conference on Computer Vision and Pattern Recognition. https://openaccess.thecvf.com/content/CVPR2023/papers/Takagi_High-Resolution_Image_Reconstruction_With_Latent_Diffusion_Models_From_Human_Brain_CVPR_2023_paper.pdf

SECTION 1

AI for neuroscience

CHAPTER ONE

AI for brain-computer interfaces

David Haslacher[a], Tugba Basaran Akmazoglu[b], Amanda van Beinum[b], Georg Starke[c,d], Maria Buthut[a], and Surjo R. Soekadar[a,*]

[a]Clinical Neurotechnology Laboratory, Department of Psychiatry and Neurosciences, Charité Campus Mitte (CCM), Charité—Universitätsmedizin Berlin, Germany
[b]Faculty of Law, University of Ottawa, Ottawa, ON, Canada
[c]Institute for History and Ethics of Medicine, School of Medicine and Health, Technical University of Munich, Germany
[d]Collège des Humanités, École Polytechnique Fédérale de Lausanne, Switzerland
*Corresponding author. e-mail address: surjo.soekadar@charite.de

Contents

1. Introduction	4
2. Basic types and applications of brain-computer interfaces	6
3. AI-enhanced brain-computer interfaces	9
3.1 Restoring sensorimotor function	10
3.2 Restoring speech	11
3.3 Treating affective disorders	13
4. Scope and scalability of AI-enhanced BCIs	14
5. Perspectives and ethical implications	16
6. Conclusions	19
Acknowledgments	20
References	20

Brain-computer interfaces (BCI) translate brain activity into control commands of external devices, for example, assistive robots or communication tools, such as smartphones. For decades, BCIs used rather simple linear classifiers, but with the availability of more powerful computers and extensive data, BCIs are now increasingly combined with advanced AI algorithms. Beyond improving neural decoding efficiency, intelligent AI-enhanced actuators or tools can also compensate for BCI's limited bandwidth, that is information transfer rates. Moreover, BCI systems that entail direct stimulation of brain tissue have become available, enabling bidirectional brain-machine interaction. Such *bidirectional* BCIs can enhance assistive applications, for example, by reinstating the feeling of touch during prosthesis control, but also pave the way for new clinical applications, for instance, termination of abnormal brain activity in epilepsy or depression. Incorporating AI-informed stimulation parameters holds the potential to enhance the richness of information conveyed from the computer to the brain. By extending the scope towards such applications, AI-enabled bidirectional BCIs may fundamentally improve treatment options for severe brain disorders, but also raise ethical questions related, for example, to mental integrity, identity, privacy, data security, accessibility, and potential misuse of neural data. Here, we provide an overview of established AI-enabled BCIs, outline current technological challenges, and discuss potential ethical implications of future advancements.

1. Introduction

Mind-reading devices were first conceptualized in the late 19th century and have since sparked the creative imagination of innovators, scientists, and artists alike. While the biological foundations of thoughts, memories and emotions were a mystery then, effects of brain stimulation hinted at the important role of electric activity in shaping these phenomena. Initial attempts to measure such activity using galvanometers (Peterson & Jung, 1907) were non-specific and imprecise. In the 1920s, however, a technique that would transform human neuroscience was discovered: Hans Berger, a psychiatrist in Jena, Germany, managed to record electric brain activity from the surface of the scalp. His technique, which became known as electroencephalography (EEG), converted the weak electric activity of the brain into visible patterns recorded on long scrolls of paper (Berger, 1929). Berger found that the EEG of an awake human was dominated by electric activity oscillating at around 10 Hz, a frequency he termed alpha. Later research found that alpha waves covaried with perception and behavior across different brain regions, for instance that occipital alpha oscillations were linked to visual function or central alpha oscillations tied to sensory and motor function. These discoveries raised the hope that, one day, deciphering EEG patterns would be possible. However, despite initial enthusiasm, it proved exceptionally challenging to map these patterns to specific thoughts, memories, or emotions. Despite its lack of specificity, EEG proved valuable in assessing a person's level of alertness and still plays an important role in diagnosing and characterizing various brain disorders, such as epilepsy (Gibbs, Davis, & Lennox, 1935), encephalitis or narcolepsy. In 1943, Grey Walter developed the first automatic EEG frequency analyzer (Walter, 1943) that was later used to provide real-time evaluation of brain activity, an important prerequisite to establish brain-computer interfaces (BCIs).

The advent of computers in the 1950s fueled expectations that the automated interpretation of electric brain signals could unlock new insights into brain functioning. In that period, Thelma Estrin, a pioneer in biomedical engineering at the University of California in Los Angeles (UCLA), successfully developed the first analog-to-digital converter, enabling the realization of an online digital signal processing system for EEG (Estrin, 1987).

Around the same time, computers also fueled the advancements of artificial intelligence (AI), that is, the ability of an artificial computational system to perform tasks that usually require human intelligence. When machines began to prove mathematical theorems and translate text between languages, it was predicted that AI would soon match human performance on any task. In 1967, Marvin Minsky, a computer scientist at the Massachusetts Institute of Technology, proclaimed that this challenge would be solved within twenty years (Mitchell, 2023). As we know now, this vision was premature but may prove to be not entirely wrong.

The dominant approach at the time is known today as "good old fashioned AI", and was based on the assumption that intelligence could be emulated by symbolic systems that used formal mathematical language to deduce facts and behaviors (Frankish & Ramsey, 2014). This paradigm was replaced in the 1980s, when a group of psychologists led by David Rumelhart showed that the chain rule of calculus, derived by Gottfried Wilhelm Leibniz in 1673, could be used to train artificial neural networks to learn arbitrary mappings between inputs and outputs (Rumelhart, Hinton, & Williams, 1986).

Parallel to AI, BCIs were gaining traction. In the early 1970s, scientists at UCLA started the first BCI project that aimed at establishing direct communication between the brain and computers using EEG signals (Vidal, 1973). This project was grounded in evidence that EEG was comprised of well-defined neural events, such as visual or auditory evoked responses, in addition to slower potentials like the *Bereitschaftspotential* (Kornhuber & Deecke, 1964; Nann, Cohen, Deecke, & Soekadar, 2019) or contingent negative variation (Walter, Cooper, Aldridge, McCallum, & Winter, 1964). It was shown that these electric potentials could be controlled by the user to establish a dialog with a computer. An important principle to establish such interaction was discovered by Eberhard Fetz: he showed that operant conditioning, a process that entails reinforcing or discouraging specific behaviors through contingent reward or punishment, was applicable to the behavior of brain cells as well (Kamiya, 1962). Due to its therapeutic potential, operant conditioning of brain activity raised widespread interest and was demonstrated both in synchronized activity of large neuronal ensembles (Sterman, Wyrwicka, & Howe, 1969) as well as single cells (Fetz, 1969).

With the availability and accessibility of computers and bio-signal amplifiers, the field of BCIs is now rapidly evolving, promising to reshape not only the way we treat brain disorders, but also how we interact with technology. In part, these advances are powered by recent developments in AI. Around 2012, when it was shown that complex neural networks could be trained to exceed human-level performance in image recognition tasks (Ciresan, Meier, Masci, & Schmidhuber, 2012; Krizhevsky, Sutskever, & Hinton, 2012), AI research and development experienced a significant resurgence of interest and investment. Termed deep learning (DL), multi-layered neural architectures were established to master challenging computational tasks such speech recognition (Zhang et al., 2018), natural language processing (Deng & Liu, 2018), car driving (Grigorescu, Trasnea, Cocias, & Macesanu, 2020), drug discovery (Chen, Engkvist, Wang, Olivecrona, & Blaschke, 2018) and finance (Heaton, Polson, & Witte, 2017). Moreover, DL has also had considerable impact on robotics (Sünderhauf et al., 2018), media recommendation (Da'u & Salim, 2020) and game playing (Mnih et al., 2013). These advances have lately also accelerated the scope and applicability of BCIs, for example to restore movement, sensation, speech, and affective well-being (Flesher et al., 2021; Hochberg et al., 2012; Metzger et al., 2023).

2. Basic types and applications of brain-computer interfaces

Depending on the purpose of use, *active*, *reactive*, and *passive* BCIs can be distinguished, each relying on specific approaches for interpreting and utilizing brain signals.

Active BCIs are designed to provide voluntary control of external devices or applications (Steinert, Bublitz, Jox, & Friedrich, 2019; Wolpaw, 2013). These BCIs translate brain activity into control commands directed at devices or tools that assist in achieving a specific intended goal. These BCIs are often used for applications such as prosthetic limb control (Soekadar et al., 2016), or communication (Birbaumer et al., 1999). The user is typically trained to operate such BCI through operant conditioning or feedback learning. During active BCI operation, the user becomes continuously informed about system performance, enabling real-time adjustments of neural activity. This feedback-mediated learning could relate to any sensory domain such as visual, haptic, tactile, proprioceptive, auditory, or even gustatory.

Instead of relying on the user's voluntary intent, *reactive* BCIs detect and interpret brain responses that occur in response to external stimulation (Farwell & Donchin, 1988; Muller-Putz, Scherer, Brauneis, & Pfurtscheller, 2005; Sellers & Donchin, 2006). Reactive BCIs commonly use predefined time windows during which brain responses to external stimuli are evaluated. These BCIs often provide higher accuracy compared to active BCIs because evoked brain responses have a high signal-to-noise ratio and known temporal characteristics. They are often used in scenarios where the timing of brain responses is crucial, such as in the selection of letters in a speller application or the detection of specific stimuli in a controlled environment.

Passive BCIs derive from neuromonitoring, for example, continuous assessment of brain responses, with the distinction that the output of passive BCIs informs the interaction with the computer or machine, typically without requiring any active user engagement (Zander & Kothe, 2011). For instance, a passive BCI could detect cognitive overload and adjust the human-computer interaction accordingly, for example in the context of air traffic controlling by reducing the number stimuli to the operator (Borghini et al., 2020; Venthur, Blankertz, Gugler, & Curio, 2010).

Another important distinction between BCI systems relates to their invasiveness (Nasr et al., 2022). Non-invasive BCIs do not require surgical intervention but typically provide lower signal to noise ratios and spatial resolution than implantable BCIs. This is because non-invasive BCIs capture brain activity from outside the skull, for example the brain's electric (Soekadar et al., 2016), magnetic (Soekadar, Witkowski, Birbaumer, & Cohen, 2015) or metabolic activity (Liew et al., 2016; Sitaram et al., 2007). The best established methods to assess the brain's metabolic activity in the context of BCIs are functional near-infrared spectroscopy (fNIRS) using light (Kohl et al., 2023; Soekadar, Kohl, Mihara, & von Lühmann, 2021) and functional magnetic resonance imaging, fMRI, using strong magnetic fields (Liew et al., 2016; Sitaram et al., 2017). While non-invasive BCIs are more susceptible to muscle activity (Muthukumaraswamy, 2013) and environmental noise (Gonzalez-Moreno et al., 2014), their accessibility, for example to allow for communication (Birbaumer et al., 1999) or control of external devices (Edelman et al., 2019; Soekadar et al., 2016; Zander, Krol, Birbaumer, & Gramann, 2016), provides a clear advantage compared to implantable BCIs.

Non-invasive active BCIs enabled patients with complete hand paralysis, for example after stroke or high spinal cord injuries, to regain lost motor functions (Soekadar, Witkowski, Vitiello, & Birbaumer, 2015). By voluntarily modulating sensorimotor rhythms detectable via EEG, patients could use these BCIs to control a robotic hand exoskeleton, restoring the ability to engage in activities of daily living (Soekadar et al., 2016). Importantly, it was shown that repeated use of such brain-controlled exoskeleton can trigger motor recovery, even in chronic paralysis (Cervera et al., 2018; Ramos-Murguialday et al., 2013), further strengthening their critical role in neurorehabilitation (Angerhöfer, Colucci, Vermehren, Hömberg, & Soekadar, 2021; Colucci et al., 2022; Simon, Bolton, Kennedy, Soekadar, & Ruddy, 2021; Ushiba & Soekadar, 2016). Beyond hand and finger control, non-invasive motor BCIs were also successfully implemented for whole-arm exoskeleton control (Crea et al., 2018). Here, a non-invasive BCI inferred the user's intention and was combined with an AI-enabled, context aware, semi-autonomous system identifying and executing the best solution via a robotic actuator, for example to grasp a glass of water and move it to the user's mouth. Besides restoring upper limb motor function, non-invasive active BCIs were also successfully used for gait restoration (He et al., 2018; Koelewijn et al., 2021), and more general to interact with an environment control interface (Bertomeu-Motos et al., 2019).

The integration of error-related potentials could further enhance such systems' accuracy and efficiency (Dornhege, Millán, Hinterberger, McFarland, & Müller, 2007). While active BCIs have shown potential in tasks such as piloting aircraft or drones (Doud, Lucas, Pisansky, & He, 2011), acquiring BCI control remains a challenging and time-consuming process that typically requires learning through operant conditioning or feedback learning. Nevertheless, applicability of such devices is currently tested in the context of entertainment systems or semi-autonomous driving (Stawicki, Gembler, & Volosyak, 2016; Wang, Li, Zhang, Li, & Zhang, 2018).

Passive BCIs, on the other hand, do not require any learning and are primarily used to improve the ergonomics of human-computer or human-machine interaction (Krol, Andreessen, & Zander, 2018; Zander & Jatzev, 2012). While such systems can improve ergonomics based on online-informed digital models built on cognitive and affective probing, that is physiological response patterns across a multitude of contexts, the decoding mechanism does not depend on the users' awareness or understanding (Krol, Haselager, & Zander, 2020) raising additional ethical concerns as compared to active BCIs.

Implantable BCIs involve a surgical procedure typically placing electrodes in direct proximity to the brain (Hochberg et al., 2012; Leuthardt, Moran, & Mullen, 2021). To better reflect different risks associated with the surgical process, they can be further distinguished in *embedded* devices that are implanted but do not enter the intracranial space, and *intracranial* devices which do (Leuthardt, Moran, & Mullen, 2021). Over the last few years, intravascular electrodes have also been established to record EEG (Oxley et al., 2021). These implanted electrodes offer signals that are less susceptible to external interference. The recorded signals may represent the aggregate electric activity of many neurons, referred to as local field potentials (LFP), or, when using microelectrodes, the electrical impulses (spikes) of individual neurons. Implantable BCIs can also allow for direct electric stimulation of brain tissue, establishing bidirectional interaction, however, with all risks associated with surgical implantation (Leuthardt et al., 2021). Consequently, the application of implantable BCIs is currently restricted to clinical applications, for example to restore movement or communication. Furthermore, these devices find utility in basic research, where they serve as valuable tools for the exploration of fundamental neural coding principles.

In the absence of powerful computers and state-of-the-art machine learning algorithms, BCIs have mainly relied on decoding approaches that operate on covariance matrices of multivariate neural data (Lotte et al., 2018). These techniques all operate under the assumption that brain data can be linearly projected to a hidden space that reveals the neural activity of interest (Haufe et al., 2014). While such linear projection may facilitate operant conditioning of neuronal cell ensembles or even singe cell activity, it is limited because it fails to capture the complexity of network interactions and nonlinear dynamics inherent in many neural processes. Therefore, application of machine learning methods, for example based on recurrent neural networks (RNN), can increase the scope and versatility of BCIs but they require a substantial amount of labeled data for effective training and might be computationally intensive, posing challenges in real-time applications and necessitating careful optimization and hardware considerations.

3. AI-enhanced brain-computer interfaces

Capitalizing on the expanded accessibility of computational capabilities and advancements in neural recording and stimulation hardware, BCIs leveraging progress in machine learning and AI hold great potential to enhance the quality of life for individuals facing severe conditions with no alternative treatments available. An example includes restoring communication in cases of complete locked-in syndrome (CLIS) (Chaudhary et al., 2022). Other clinical applications are conceivable, for example to restore cognitive or affective function. The integration of AI-enhanced electric brain stimulation into these systems, allowing for bidirectional interaction between the brain and computer, is particularly desirable because it can improve efficacy of BCIs. However, it also requires specific considerations due to potential AI-related biases or misconfigurations that could impact the reliability, safety, and fairness of the system.

3.1 Restoring sensorimotor function

Motor BCIs were originally designed to decode movement-related signals from the brain translating these signals into motor commands controlling a robotic prosthesis. In 2012, it was demonstrated that after years of severe paralysis, a tetraplegic patient could successfully operate a robotic arm in three dimensions using an implantable neural interface (Hochberg et al., 2012). In another study, a tetraplegic person was able to perform a range of wrist and hand movements based on intracortical recordings translated into commands of a custom-built high-resolution neuromuscular electrical stimulation (Bouton et al., 2016).

An important advancement of such systems was also achieved when sensory feedback could be re-established based on the conversion of external stimuli, such as pressure or temperature on a prosthetic limb, into direct electric stimulation of brain tissue. This can be achieved via electrodes implanted into cortical regions linked to the targeted sensory function but optimal implantation sites and stimulation parameters are not yet well established. Typically, such artificial sensory feedback is dynamically adjusted in response to changing environmental conditions. Thereby, researchers demonstrated that BCI-based control of external actuators could be enhanced by transmitting artificial tactile feedback using intracortical microstimulation of the somatosensory cortex in non-human primates (O'Doherty et al., 2011). This technology holds the potential to enhance clinical motor neuroprostheses by creating artificial somatic perceptions linked to prosthetic limbs or virtual objects. These results were successfully extended to humans, and it was shown that BCI-based prosthetic arm control could be enhanced by direct electric stimulation of the somatosensory cortex of a tetraplegic patient (Flesher et al., 2021).

Remarkably, electric stimulation of the brain may be useful even when not designed to mimic natural neural codes. For instance, non-human primates could learn to use an artificial electric signal applied to the cortex to perceive hand positions relative to an unseen target. By combining it with vision, primates could form an optimal estimate of hand position (Dadarlat, O'Doherty, & Sabes, 2015). Critically, the stimulation was intentionally designed not to resemble neural codes for these signals. These results show that the brain can learn to use signals from a BCI with an initially unfamiliar encoding, with potential applications in restoring proprioception and studying mechanisms underlying sensory integration.

In case peripheral nerves are intact, versatility of prosthetic control can also be restored using cuff and intraneural electrodes, for example attached to the median nerve (Zollo et al., 2019). Such peripheral neural feedback can significantly enhance the ability to grasp and manipulate objects for complex tasks. To optimize electric stimulation protocols of nerves, an artificial model of the brain could be used. For instance, an artificial model of the visual system was designed to optimize electric stimulation parameters to elicit specific visual percepts during electric stimulation of the optic nerve (Haslacher, Nasr, & Soekadar, 2021; Romeni, Zoccolan, & Micera, 2021). Some sensory modalities can even be effectively replaced without an invasive approach. For instance, thermal stimuli applied to the residual limb could restore thermal sensation in the phantom limb of amputees (Iberite et al., 2023).

Complementing Walter Penfield's notion of the topographic organization of the body in the motor cortex proposed nearly 100 years ago (Penfield & Boldrey, 1937), it is now widely accepted that this topographic map is interspersed with networks that integrate motor effectors with sensory information and other brain functions (Gordon et al., 2023). Remarkably, a simple linear regression model can successfully map neuronal firing rates recorded using electrode arrays implanted in the hand and arm regions of the motor cortex to five-dimensional movement trajectories of a robotic exoskeleton (Flesher et al., 2021). Most commonly, though, linear or nonlinear variants of the Kalman filter are employed to decode such motion trajectories (Dong et al., 2023). In BCIs that restore sensorimotor function, decoding strategies building on DL approaches, for example RNNs, are currently substantially less common than in BCIs that restore speech.

3.2 Restoring speech

The first clinically meaningful BCI application enabled patients with locked-in syndrome (LIS) to select letters on a screen by modulating slow cortical potentials (Birbaumer et al., 1999). Later demonstrations using SSVEP BCIs allowed for faster spelling, but did not reach communications speeds achievable by other means, for example muscle or eye movements recorded by electromyography (EMG) or electrooculography (EOG). That is one of the main reasons why BCIs did not find their way into broader clinical adoption in LIS. For many decades and despite major research efforts, restoration of communication in CLIS, that is once any form of EMG or EOG seized, failed. There were various hypotheses, for example the extinction of thought hypothesis (Birbaumer, Gallegos-Ayala, Wildgruber, Silvoni, & Soekadar, 2014) or fragmentation of sleep (Soekadar et al., 2013), to explain the inability to reinstate communication in CLIS. After some promising but also controversial studies using fNIRS (Chaudhary, Xia, Silvoni, Cohen, & Birbaumer, 2017; Gallegos-Ayala et al., 2014), a study using an implantable BCI convincingly demonstrated successful communication in CLIS in a young man who modulated neural firing rates based on auditory feedback to select letters one at a time to form words and phrases (Chaudhary et al., 2022). This study showed that—in principle—communication in CLIS is possible.

Another BCI approach to establish communication aims at speech decoding using implantable electrodes. Here, the structure of language is an essential factor to establish such BCIs systems. To leverage prior knowledge on language structure in the decoding of linguistic content from dynamic brain activity, sophisticated modeling approaches from natural language processing are required. The first step of such decoding pipelines consists of the extraction of simple phonetic elements, such as phonemes or phones, from cortical areas involved in speech production and perception (Anumanchipalli, Chartier, & Chang, 2019; Makin, Moses, & Chang, 2020; Moses, Leonard, Makin, & Chang, 2019). Alternatively, letters can be extracted from neural recordings in the motor cortex during imagined handwriting (Willett, Avansino, Hochberg, Henderson, & Shenoy, 2021). This is typically achieved using RNNs. Subsequently, these linguistic elements are combined to form words. The latter step is guided by large-vocabulary probabilistic models that infer the most likely word imagined by the user (Willett et al., 2023). Furthermore, probabilities of individual words may be inferred from their context in a sentence using large language models such as GPT-1, the forerunner of the model behind ChatGPT (Tang, LeBel, Jain, & Huth, 2023).

In a study pursued at the University of California in San Francisco, ECoG signals were harnessed to decode both spoken responses and perceived questions with accuracy rates reaching levels of up to 76% for perceived questions and 61% for produced answers (Moses et al., 2019). Another study used RNN to decode cortical activity into articulatory movement representations, and then transformed those representations into speech acoustics (Anumanchipalli et al., 2019). The decoder could also synthesize speech when a participant silently mimed sentences. Further improvements in the decoder pipeline resulted in word error rates as low as 3% when constrained to a limited vocabulary (Makin et al., 2020). Aiming at large-vocabulary decoding, extensive training of deep-learning models using neural data collected during attempted and silently spoken sentences could achieve a median rate of 78 words per minute and median word error rate of 25% (Metzger et al., 2023).

Besides ECoG, intracortical microelectrode arrays measuring neuronal spiking activity have also been used successfully for speech decoding. Word error rates as low as 9.1% to 23.8% on a very large vocabulary (125,000 words) were achieved (Willett et al., 2023). With such bandwidth or ITR, these systems outrun other means for communication in LIS and hold great potential to improve quality of life in severe paralysis. However, medical risks must be evaluated on an individual level and might be prohibitive to proceed with implantation, and it is unclear whether these results can be generalized to larger and heterogenous patient populations. Furthermore, it is unclear whether such an approach would be feasible in CLIS patients or in patients transitioning into CLIS.

3.3 Treating affective disorders

Whereas motor BCIs typically build on cortical interfaces, brain–computer interaction with subcortical brain areas is also possible. For such purpose, particularly fMRI has been broadly explored (Liew et al., 2016; Sitaram et al., 2007), but also deep brain stimulation (DBS) can be used (Benabid et al., 2011). Beyond the motor domain, for instance, in the treatment of Parkinson's (Swann et al., 2018), also affective disorders, such as severe treatment-resistant depression, can be targeted. In this context, multi-day intracranial electrophysiology helped to identify symptom-specific biomarkers (Scangos, Makhoul, Sugrue, Chang, & Krystal, 2021). In a next step, a BCI consisting of two electrode leads placed in subcortical brain structures could reliably disrupt brain dynamics linked to symptom aggravation whenever such biomarker was detected (Scangos, Khambhati, et al., 2021). The advantage of being able to record and stimulate at the same time was also demonstrated by a study that used DBS of the sub-callosal cingulate. By applying an explainable AI approach, a link between LFP changes and the patient's current clinical state could be identified (Alagapan et al., 2023).

To capture symptom-specific biomarkers of depression, multi-modal approaches have proven useful. Electrophysiological recordings were first processed using conventional signal processing techniques to obtain power, coherence, and phase-amplitude coupling, reflecting the size of and relationships between neural oscillations of specific frequencies. These features were then classified as normal or pathological using neural networks or logistic regression (Alagapan et al., 2023; Scangos, Khambhati, et al., 2021; Scangos, Makhoul, et al., 2021). Importantly, a recent study has made use of sophisticated image processing techniques such as convolutional neural networks to reveal changes in facial expressions that track both depression symptoms and cingulate cortex activity (Alagapan et al., 2023). It is likely that such approach might also generalize beyond affective disorders and allow for data-driven predictions and characterizations of a wider range of affective and cognitive states or symptoms.

4. Scope and scalability of AI-enhanced BCIs

As highlighted in the previous section, AI algorithms can improve performance of both implantable and non-invasive BCIs. However, BCIs that require surgical procedures face significant challenges when it comes to scalability due to medical risks associated with implantation (Leuthardt et al., 2021). These risks include, besides the general risk of anesthesia, severe infections, tissue damage and bleeding. Moreover, repair or replacement requires another surgical intervention. Taking such medical risks might be justified in severe conditions, for example LIS or CLIS, but seem prohibitive in healthy populations. In contrast, non-invasive BCIs do not pose any of these medical risks, but—depending on the desired application—struggle to achieve the necessary focality, depth, and signal quality required for precise and reliable brain recordings and closed-loop stimulation (Nasr et al., 2022). However, advancements in recording and stimulation technology are actively addressing these shortcomings, with recent research enhancing the capabilities of non-invasive BCIs. By combining these advances with AI-enhanced, context-sensitive tools that compensate for limited BCI bandwidth, necessity for surgical interventions may diminish, paving the way for a more accessible and user-friendly integration.

While for example EEG provides limited signal-to-noise ratios, especially in higher frequency bands (Muthukumaraswamy, 2013), optically pumped magnetometers (OPM) or other quantum sensors that work at near room-temperature provide much better signals and spatial resolution (Brookes et al., 2022). In contrast to EEG, OPMs are capable of accurately measuring oscillatory brain activity passing through the skull without distortion. Given appropriate calibration, OPMs could potentially achieve a higher spatial resolution than any existing non-invasive tool to assess oscillatory brain activity (Iivanainen, Stenroos, & Parkkonen, 2017). First steps to use OPMs for BCI applications were successfully established (Zerfowski, Sander, Tangermann, Soekadar, & Middelmann, 2021), but more studies are needed to identify applications for which OPMs are most effectively suited.

While DBS showed impressive capabilities to alleviate clinical symptoms and to modulate brain activity, surgical risks and inflexibility regarding the stimulation site limit its scalability. While more scalable, non-invasive brain stimulation tools like transcranial electric or magnetic stimulation, provide only low spatial resolution and cannot effectively target deep brain regions (Nasr et al., 2022). Temporally interfering magnetic stimulation (TIMS), in contrast, addresses these challenges by applying two high-frequency magnetic fields through a pair of coils, creating an interference pattern that is locally amplitude-modulated to affect brain activity only in a small region (Nasr, Haslacher, & Soekadar, 2023; Soekadar & Nasr, 2023). TIMS offers higher spatial resolution compared to traditional methods and the ability to selectively influence deep brain regions without affecting superficial areas. Combined with stimulation artifact rejection algorithms developed for EEG/MEG (Haslacher et al., 2023; Haslacher, Nasr, Robinson, Braun, & Soekadar, 2021), TIMS could be adapted to brain activity in real-time to deliver non-invasive closed-loop neuromodulation. Another non-invasive approach directly compatible with EEG/MEG is transcranial focused ultrasound stimulation, although distortion of the acoustic field by the skull remains a challenge (Folloni et al., 2019). Overall, the use of AI algorithms is advancing the scope, performance, usability, and practicality of BCI systems. Besides contributing to decoding neural activity, AI can be also instrumental to compensate for limited BCI information transfer rates, for example by introducing context sensitivity (Catalán et al., 2023; Crea et al., 2018). The development and integration of quantum computer-based AI (Quantum AI) hold the promise of further catalyzing ongoing advancements but also bring about new challenges and possibly uncharted ethical issues in the context of human-computer interaction.

5. Perspectives and ethical implications

While both BCI technology (Clausen et al., 2017) and AI pose various ethical challenges, the question arises: does their combination raise any novel and unique ethical challenges?

While a comprehensive answer is beyond the scope of this chapter, the body of literature dedicated to the ethics of AI in medicine in general (Char, Shah, & Magnus, 2018; Vayena, Blasimme, & Cohen, 2018) and to AI in clinical neuroscience and psychiatry in particular (Friedrich, Wolkenstein, Bublitz, Jox, & Racine, 2021; Ienca & Ignatiadis, 2020; Jotterand & Ienca, 2021; Starke, De Clercq, Borgwardt, & Elger, 2021) can provide some guidance and structure for addressing this question. For instance, a review of 84 international ethics guidelines identified five key principles as the most common denominators defining the global landscape of ethical guidance, namely privacy, justice and fairness, responsibility, transparency, and non-maleficence (Jobin, Ienca, & Vayena, 2019), a finding largely concurrent with similar research (Fjeld, Achten, Hilligoss, Nagy, & Srikumar, 2020). Looking at AI-enhanced BCIs by drawing on these shared principles allows a systematic approach to surveying ethical challenges at the intersection of novel neurotechnologies and AI.

For the successful implementation of self-learning AI algorithms for neural decoding, significant neural data must be recorded and stored. This data is used initially to train the model and subsequently to guide its functionality, encompassing the read-out and write-in capabilities of the device. A critical ethical consideration related to AI in this context is therefore the *security* and *privacy* of the transmitted data among servers, clinics, and devices. The protection of this information is crucial, given its sensitive nature. Furthermore, the usage of data from implanted devices, such as pacemakers, as valid evidence in legal cases (Dobson, 2003) introduces new vulnerabilities for users of implanted AI-enhanced BCI devices. To reduce data collection times in each BCI that can be extensive (Berezutskaya et al., 2023) and tiring (Thielen, Marsman, Farquhar, & Desain, 2021), there are efforts to train algorithmic models via transfer learning from able-bodied participants (Berezutskaya et al., 2023) and to implement "subject-independent" BCIs based on deep convolutional neural networks (Kwon, Lee, Guan, & Lee, 2020).

Given the rapid progress in technology, it may therefore not be long before algorithmic components of BCIs are trained on data collected from larger groups rather than on individual neural data sets. While this development would likely improve the technology's ease of use and make BCIs accessible to larger groups of patients, this scenario also raises ethical concerns related to the *fairness* and *justice*. The dangers of algorithmic bias in medical AI are well documented as is the possibility that algorithms may prioritize efficiency over concerns with fairness (Chen et al., 2023; Obermeyer, Powers, Vogeli, & Mullainathan, 2019). These difficulties could be exacerbated in clinical neurotechnology by notorious difficulties of defining appropriate ground truths when training medical algorithms (Starke, De Clercq, & Elger, 2021) and by the almost exclusive development of such AI-based systems in high-income countries (Weissglass, 2022). Such global imbalances and injustices also carry the potential risk of commodifying neural data and trading such sensitive data in data markets unless required security, privacy arrangements and policies are in place. Therefore, global synergistic actions deployed through co-design, co-implementation and collaborative governance that create equitable and inclusive access to services fostering brain health are critical (Owolabi et al., 2023).

A third major ethical concern relates to questions of *responsibility* when AI models become involved in the decision-making processes of the user, with potential impact on their agency and autonomy (Sand, Durán, & Jongsma, 2022). Ideally, AI should serve the optimal execution of the user's intentions or will, while not interfering with the user's authorship of action. Though there are medical conditions where the AI component of BCI devices may contribute to returning control over the user's bodily or mental functions, for example detecting and preventing epileptic seizures, overreliance on these devices may lead to harm when devices malfunction (Gilbert, 2015). In cases where AI-informed processes directly influence user intentions or will, for example with implanted devices modifying substance use or other behaviors, several ethical concerns arise regarding the potential manipulation of fundamental *human agency* and the erosion of *personal autonomy*. While there may be medical conditions in which such applications cannot be denied categorically, for example people suffering from severe forms of psychosis, addiction, or impulsivity, there are compelling reasons for a ban of such systems for as long as there is lack of any regulatory and technological mechanisms to protect human autonomy and agency.

The delicate balance between technological advancement and safeguarding individual sovereignty requires vigilant consideration of the far-reaching implications of AI-enabled BCI systems. In this context, *transparency* is of particular importance, offering some level of explainability and interpretability of the decoding processes in these devices. These forms of algorithmic transparency and explainability need to be tailored to the needs of different involved groups (Arbelaez Ossa et al., 2022) and should be critically assessed and measured for their quality (Holzinger, Langs, Denk, Zatloukal, & Müller, 2019). For some BCIs, for example closed-loop devices such as brain-responsive neurostimulation, the lack of explanation of individual decisions may also provide grounds to prefer locked algorithms that can be assessed more easily before their deployment (Liao, 2023).

Finally, as always in medicine, questions of non-maleficence need to be considered carefully. In the context of speech BCIs, if the algorithm is unable to accurately interpret the intended speech, or in the case of motor BCIs, if it fails to correctly interpret intended movements, the resulting output of the human-machine co-processing may be inaccurate. Besides threats to autonomy, control and agency, such errors could also have legal consequences when unintended words or actions lead to external harms (Chandler et al., 2022). Hence accuracy and robustness, or more aptly, the lack thereof, are a significant ethical concern when BCI devices are integrated with AI. Applying rigorous standards to AI-enhanced BCIs throughout the product lifecycle starting from the design process to post-market surveillance efforts is therefore a key tenet of responsible engineering and governance. Implementing comprehensive regulatory frameworks and robust oversight mechanisms is therefore imperative to mitigate the ethical ramifications of AI-enabled, bidirectional BCI, to foster public trust and ensure accountability among stakeholders. Research is needed that establishes adequate tools to investigate the bio-psycho-social impact of AI-enhanced BCIs and their impact on society in a long-term perspective (>10 years) (Soekadar et al., 2023).

While the ethical concerns related to AI-enhanced BCIs outlined here overlap with concerns with AI systems in general, they are - in some ways - particular due to the unique nature of direct interaction with the human brain and the capabilities of these novel tools to reshape their mode of being-in-the-world ('In-der-Welt-sein'). Also from a legal point of view, the inclusion of AI in BCI systems raises additional questions about categorizing devices that directly interface with the brain—whether they should be treated as material objects or body parts—and how their explantation should be governed (Bublitz & Gilbert, 2023). The incorporation of AI components not only facilitates bidirectional functionalities but, in some instances, has the potential to restore previously lost bodily functions such as speech, grasp, and sensory perception. In these circumstances, the relationships between user and device may be closer to embodied integration rather than legal property ownership. The classification of AI-enabled BCI devices as inert objects or bodily components holds significant implications for the expectations of users, governments, developers, and third parties, as well as for legal responses. Therefore, special caution is essential in making decisions about how these devices are categorized. The metaphors used to describe and introduce AI-enabled BCI technologies have a similar far-reaching potential to influence user and regulatory expectations (Richards & Smart, 2013). While metaphors can be useful tools to frame the benefits of this new technology (Leta Jones & Millar, 2017), use of inappropriate or suggestive metaphors such as "remote controlled minds", "your brain with super powers" or new human "spare parts" may contribute to the generation of hype and fear, misinform user groups, and cause unexpected legal responses. Intentional or unintentional negative framing may even undermine product development and impede the successful communication of a product's benefits and features. It can lead to a skewed perception among users, hinder adoption rates, and create unnecessary challenges for marketing efforts. Therefore, careful consideration of language and framing is crucial to ensure a positive and accurate portrayal of AI-enhanced BCIs.

To move towards these desiderata, it is crucial to involve both experts from various fields, such as neuroethicists, psychologists, sociologists, clinicians, policymakers, lawyers, as well as device users, to clearly define the acceptable boundaries of use and prevent the potential exploitation of vulnerable populations. Identifying and understanding stakeholders is a fundamental step in the design process, with particular emphasis on directly affected stakeholders like device users. In this regard, the value-sensitive design approach (Friedman & Kahn, 2002; Friedman, Kahn, & Borning, 2002), widely accepted in the AI ethics community, proves to be a valuable tool for crafting these innovative technologies. The value-sensitive design approach integrates participatory design, democratic involvement, and human values into the design process. It involves examining and weighing value trade-offs against each other. Adopting a user-centric stance (Ienca, Kressig, Jotterand, & Elger, 2017), this approach prioritizes user experience, addressing their fears, aspirations, and complaints throughout the process. Specific suggestions how to achieve value-sensitive and user-centric design of AI-based neuro-technology using embedded ethics or community-led approaches to value alignment are provided in later chapters in this volume.

6. Conclusions

The most advanced BCI applications today can modulate, assist, support, augment or restore a diverse range of functions including movement, sensation, speech, and affective well-being. In such applications, interdependencies may emerge between the user and computer, and these interdependencies tend to intensify as AI components become more deeply integrated. While the computer learns from brain signals through statistical analysis, the brain learns from sensory feedback or direct, closed-loop stimulation as outputs of the computer. Such direct stimulation comes with the risk of taking the individual agent "out of the loop" of control (Goering, Klein, Dougherty, & Widge, 2017) particularly in domains where conscious and voluntary control represents a fundamental aspect of autonomy.

Since current bidirectional BCIs typically make use of invasive recording and stimulation techniques, efforts and risks associated with implantation have prevented widespread adoption of these technologies. Advances in non-invasive recording and stimulation techniques may foster broader adoption of medical but also non-medical BCI applications. As these technologies are still in the early stages of development, it is both cost-effective and strategically wise to incorporate essential measures ensuring human value-alignment right from their initial design phases. Including considerations of AI ethics principles early in the convergence with BCI technology serves as a preventive measure, minimizing potential harms of such technology to users and society at present and in the future.

Acknowledgments

This chapter and the presented studies were supported by the ERA-NET NEURON project HYBRIDMIND (BMBF, 01GP2121B), the European Research Council (ERC) under the project NGBMI (759370) and TIMS (101081905), the Collaboration of Humanities and Social Sciences in Europe (CHANSE) under the project Q-SHIFT (01UX2211), the Federal Ministry of Research and Education (BMBF) under the projects SSMART (01DR21025A), NEO (13GW0483C), QHMI (03ZU1110DD) and NeuroQ (13N16486), and the Einstein Foundation Berlin (ESB).

References

Alagapan, S., Choi, K. S., Heisig, S., Riva-Posse, P., Crowell, A., Tiruvadi, V., ... Rozell, C. J. (2023). Cingulate dynamics track depression recovery with deep brain stimulation. *Nature*. https://doi.org/10.1038/s41586-023-06541-3.

Angerhöfer, C., Colucci, A., Vermehren, M., Hömberg, V., & Soekadar, S. R. (2021). Post-stroke rehabilitation of severe upper limb paresis in Germany—Toward long-term treatment with brain-computer interfaces [Mini review]. *Frontiers in Neurology, 12*(2166), https://doi.org/10.3389/fneur.2021.772199.

Anumanchipalli, G. K., Chartier, J., & Chang, E. F. (2019). Speech synthesis from neural decoding of spoken sentences. *Nature, 568*(7753), 493–498. https://doi.org/10.1038/s41586-019-1119-1.

Arbelaez Ossa, L., Starke, G., Lorenzini, G., Vogt, J. E., Shaw, D. M., & Elger, B. S. (2022). Re-focusing explainability in medicine. *Digital Health, 8*, 20552076221074488.

Benabid, A. L., Costecalde, T., Torres, N., Moro, C., Aksenova, T., Eliseyev, A., ... Chabardes, S. (2011). Deep brain stimulation: BCI at large, where are we going to? *Progress in Brain Research, 194*, 71–82. https://doi.org/10.1016/b978-0-444-53815-4.00016-9.

Berezutskaya, J., Freudenburg, Z. V., Vansteensel, M. J., Aarnoutse, E. J., Ramsey, N. F., & van Gerven, M. A. J. (2023). Direct speech reconstruction from sensorimotor brain activity with optimized deep learning models. *Journal of Neural Engineering, 20*(5), https://doi.org/10.1088/1741-2552/ace8be.

Berger, H. (1929). Über das Elektrenkephalogramm des Menschen. *Archiv für Psychiatrie und Nervenkrankheiten, 87*(1), 527–570. https://doi.org/10.1007/BF01797193.

Bertomeu-Motos, A., Ezquerro, S., Barios, J. A., Lledó, L. D., Domingo, S., Nann, M., ... Garcia-Aracil, N. (2019). User activity recognition system to improve the performance of environmental control interfaces: A pilot study with patients. *Journal of Neuroengineering and Rehabilitation, 16*(1), 10. https://doi.org/10.1186/s12984-018-0477-5.

Birbaumer, N., Gallegos-Ayala, G., Wildgruber, M., Silvoni, S., & Soekadar, S. R. (2014). Direct brain control and communication in paralysis. *Brain Topography, 27*(1), 4–11. https://doi.org/10.1007/s10548-013-0282-1.

Birbaumer, N., Ghanayim, N., Hinterberger, T., Iversen, I., Kotchoubey, B., Kubler, A., ... Flor, H. (1999). A spelling device for the paralysed. *Nature, 398*(6725), 297–298. https://doi.org/10.1038/18581.

Borghini, G., Di Flumeri, G., Aricò, P., Sciaraffa, N., Bonelli, S., Ragosta, M., ... Babiloni, F. (2020). A multimodal and signals fusion approach for assessing the impact of stressful events on air traffic controllers. *Scientific Reports, 10*(1), 8600. https://doi.org/10.1038/s41598-020-65610-z.

Bouton, C. E., Shaikhouni, A., Annetta, N. V., Bockbrader, M. A., Friedenberg, D. A., Nielson, D. M., ... Mysiw, W. J. (2016). Restoring cortical control of functional movement in a human with quadriplegia. *Nature, 533*(7602), 247–250.

Brookes, M. J., Leggett, J., Rea, M., Hill, R. M., Holmes, N., Boto, E., & Bowtell, R. (2022). Magnetoencephalography with optically pumped magnetometers (OPM-MEG): The next generation of functional neuroimaging. *Trends in Neurosciences, 45*(8), 621–634. https://doi.org/10.1016/j.tins.2022.05.008.

Bublitz, J. C., & Gilbert, F. (2023). Legal aspects of unwanted device explantations: A comment on the patient R case. *Brain Stimulation, 16*(5), 1425–1429.

Catalán, J. M., Trigili, E., Nann, M., Blanco-Ivorra, A., Lauretti, C., Cordella, F., ... García-Aracil, N. (2023). Hybrid brain/neural interface and autonomous vision-guided whole-arm exoskeleton control to perform activities of daily living (ADLs). *Journal of Neuroengineering and Rehabilitation, 20*(1), 61. https://doi.org/10.1186/s12984-023-01185-w.

Cervera, M. A., Soekadar, S. R., Ushiba, J., Millan, J. D. R., Liu, M., Birbaumer, N., & Garipelli, G. (2018). Brain-computer interfaces for post-stroke motor rehabilitation: A meta-analysis. *Annals of Clinical and Translational Neurology, 5*(5), 651–663. https://doi.org/10.1002/acn3.544.

Chandler, J., van der Loos, K., Boehnke, S., Buchman, D. Z., Beaudry, J. S., & Illes, J. (2022). Brain computer interfaces and communication disabilities: Ethical, legal and social aspects of decoding speech from the brain. *Frontiers in Human Neuroscience (in press)*.

Char, D. S., Shah, N. H., & Magnus, D. (2018). Implementing machine learning in health care—Addressing ethical challenges. *The New England Journal of Medicine, 378*(11), 981.

Chaudhary, U., Vlachos, I., Zimmermann, J. B., Espinosa, A., Tonin, A., Jramillo-Gonzales, A., ... Birbaumer, N. (2022). Spelling interface using intracortical signals in a completely locked-in patient enabled via auditory neurofeedback training. *Nature Communications* (in press).

Chaudhary, U., Xia, B., Silvoni, S., Cohen, L. G., & Birbaumer, N. (2017). Brain-computer interface-based communication in the completely locked-in state. *PLoS Biology, 15*(1), e1002593. https://doi.org/10.1371/journal.pbio.1002593.

Chen, H., Engkvist, O., Wang, Y., Olivecrona, M., & Blaschke, T. (2018). The rise of deep learning in drug discovery. *Drug Discovery Today, 23*(6), 1241–1250.

Chen, R. J., Wang, J. J., Williamson, D. F., Chen, T. Y., Lipkova, J., Lu, M. Y., ... Mahmood, F. (2023). Algorithmic fairness in artificial intelligence for medicine and healthcare. *Nature Biomedical Engineering, 7*(6), 719–742.

Ciresan, D., Meier, U., Masci, J., & Schmidhuber, J. (2012). Multi-column deep neural network for traffic sign classification. *Neural Networks: The Official Journal of the International Neural Network Society, 32*, 333–338. https://doi.org/10.1016/j.neunet.2012.02.023.

Clausen, J., Fetz, E., Donoghue, J., Ushiba, J., Sporhase, U., Chandler, J., ... Soekadar, S. R. (2017). Help, hope, and hype: Ethical dimensions of neuroprosthetics. *Science (New York, N. Y.), 356*(6345), 1338–1339. https://doi.org/10.1126/science.aam7731.

Colucci, A., Vermehren, M., Cavallo, A., Angerhöfer, C., Peekhaus, N., Zollo, L., ... Soekadar, S. R. (2022). Brain-computer interface-controlled exoskeletons in clinical neurorehabilitation: Ready or not? *Neurorehabilitation and Neural Repair, 36*(12), 747–756. https://doi.org/10.1177/15459683221138751.

Crea, S., Nann, M., Trigili, E., Cordella, F., Baldoni, A., Badesa, F. J., ... Soekadar, S. R. (2018). Feasibility and safety of shared EEG/EOG and vision-guided autonomous whole-arm exoskeleton control to perform activities of daily living. *Scientific Reports, 8*(1), 10823. https://doi.org/10.1038/s41598-018-29091-5.

Da'u, A., & Salim, N. (2020). Recommendation system based on deep learning methods: A systematic review and new directions. *Artificial Intelligence Review, 53*(4), 2709–2748.

Dadarlat, M. C., O'Doherty, J. E., & Sabes, P. N. (2015). A learning-based approach to artificial sensory feedback leads to optimal integration. *Nature Neuroscience, 18*(1), 138–144. https://doi.org/10.1038/nn.3883.

Deng, L., & Liu, Y. (2018). *Deep learning in natural language processing*. Springer.

Dobson, R. (2003). Pacemaker pinpoints time of death in murder case. *BMJ (Clinical Research Ed.), 326*(7380), 70.

Dong, Y., Wang, S., Huang, Q., Berg, R. W., Li, G., & He, J. (2023). Neural decoding for intracortical brain-computer interfaces. *Cyborg and Bionic Systems, 4*, 0044. https://doi.org/10.34133/cbsystems.0044.

Dornhege, G., Millán, J. d R., Hinterberger, T., McFarland, D. J., & Müller, K.-R. (2007). *Error-related EEG potentials in brain-computer interfaces. Toward Brain-Computer Interfacing.* MIT Press,291–301. http://ieeexplore.ieee.org/document/6281216.

Doud, A. J., Lucas, J. P., Pisansky, M. T., & He, B. (2011). Continuous three-dimensional control of a virtual helicopter using a motor imagery based brain-computer interface. *PLoS One, 6*(10), e26322. https://doi.org/10.1371/journal.pone.0026322.

Edelman, B. J., Meng, J., Suma, D., Zurn, C., Nagarajan, E., Baxter, B. S., ... He, B. (2019). Noninvasive neuroimaging enhances continuous neural tracking for robotic device control. *Science Robotics, 4*(31), https://doi.org/10.1126/scirobotics.aaw6844.

Estrin, T. (1987). The UCLA Brain Research Institute data processing laboratory. In *Proceedings of ACM conference on History of medical informatics*. Bethesda, MD. https://doi.org/10.1145/41526.41533.

Farwell, L. A., & Donchin, E. (1988). Talking off the top of your head: Toward a mental prosthesis utilizing event-related brain potentials. *Electroencephalography and Clinical Neurophysiology, 70*(6), 510–523. https://doi.org/10.1016/0013-4694(88)90149-6.

Fetz, E. E. (1969). Operant conditioning of cortical unit activity. *Science (New York, N. Y.), 163*(3870), 955–958. https://doi.org/10.1126/science.163.3870.955.

Fjeld, J., Achten, N., Hilligoss, H., Nagy, A., & Srikumar, M. (2020). Principled artificial intelligence: Mapping consensus in ethical and rights-based approaches to principles for AI. *SSRN Electronic Journal*. https://doi.org/10.2139/ssrn.3518482.

Flesher, S. N., Downey, J. E., Weiss, J. M., Hughes, C. L., Herrera, A. J., Tyler-Kabara, E. C., ... Gaunt, R. A. (2021). A brain-computer interface that evokes tactile sensations improves robotic arm control. *Science (New York, N. Y.), 372*(6544), 831–836. https://doi.org/10.1126/science.abd0380.

Folloni, D., Verhagen, L., Mars, R. B., Fouragnan, E., Constans, C., Aubry, J. F., ... Sallet, J. (2019). Manipulation of subcortical and deep cortical activity in the primate brain using transcranial focused ultrasound stimulation. *Neuron, 101*(6), 1109–1116.e1105. https://doi.org/10.1016/j.neuron.2019.01.019.

Frankish, K., & Ramsey, W. M. (2014). *The Cambridge handbook of artificial intelligence.* Cambridge University Press. (Cover image). http://assets.cambridge.org/97805218/71426/cover/9780521871426.jpg.

Friedman, B., P. H. Kahn, Jr., and A. Borning. 2002. Value Sensitive Design: Theory and Methods (UW CSE Technical Report 02-12-01). ftp://ftp.cs.washington.edu/tr/2002/12/UW-CSE-02-12-01.pdf. https://faculty.washington.edu/pkahn/articles/vsd-theory-methods-tr.pdf.

Friedman, B., & Kahn, P. H. (2002). *Human values, ethics, and design. The human-computer interaction handbook: Fundamentals, evolving technologies and emerging applications.* L. Erlbaum Associates Inc, 1177–1201.

Friedrich, O., Wolkenstein, A., Bublitz, C., Jox, R. J., & Racine, E. (2021). *Clinical neurotechnology meets artificial intelligence.* Springer.

Gallegos-Ayala, G., Furdea, A., Takano, K., Ruf, C. A., Flor, H., & Birbaumer, N. (2014). Brain communication in a completely locked-in patient using bedside near-infrared spectroscopy. *Neurology, 82*(21), 1930–1932. https://doi.org/10.1212/WNL.0000000000000449.

Gibbs, F. A., Davis, H., & Lennox, W. G. (1935). The electro-encephalogram in epilepsy and in conditions of impaired consciousness. *Archives of Neurology & Psychiatry, 34*(6), 1133–1148. https://doi.org/10.1001/archneurpsyc.1935.02250240002001.

Gilbert, F. (2015). A threat to autonomy? The intrusion of predictive brain implants. *Ajob Neuroscience, 6*(4), 4–11.

Goering, S., Klein, E., Dougherty, D. D., & Widge, A. S. (2017). Staying in the loop: Relational agency and identity in next-generation DBS for psychiatry. *Ajob Neuroscience, 8*(2), 59–70. https://doi.org/10.1080/21507740.2017.1320320.

Gonzalez-Moreno, A., Aurtenetxe, S., Lopez-Garcia, M. E., del Pozo, F., Maestu, F., & Nevado, A. (2014). Signal-to-noise ratio of the MEG signal after preprocessing. *Journal of Neuroscience Methods, 222*, 56–61. https://doi.org/10.1016/j.jneumeth.2013.10.019.

Gordon, E. M., Chauvin, R. J., Van, A. N., Rajesh, A., Nielsen, A., Newbold, D. J., ... Dosenbach, N. U. F. (2023). A somato-cognitive action network alternates with effector regions in motor cortex. *Nature, 617*(7960), 351–359. https://doi.org/10.1038/s41586-023-05964-2.

Grigorescu, S., Trasnea, B., Cocias, T., & Macesanu, G. (2020). A survey of deep learning techniques for autonomous driving. *Journal of Field Robotics, 37*(3), 362–386.

Haslacher, D., Narang, A., Sokoliuk, R., Cavallo, A., Reber, P., Nasr, K., ... Soekadar, S. R. (2023). In vivo phase-dependent enhancement and suppression of human brain oscillations by transcranial alternating current stimulation (tACS). *Neuroimage, 275*, 120187. https://doi.org/10.1016/j.neuroimage.2023.120187.

Haslacher, D., Nasr, K., Robinson, S. E., Braun, C., & Soekadar, S. R. (2021). Stimulation Artifact Source Separation (SASS) for assessing electric brain oscillations during transcranial alternating current stimulation (tACS). *Neuroimage* 117571. https://doi.org/10.1016/j.neuroimage.2020.117571.

Haslacher, D., Nasr, K., & Soekadar, S. R. (2021). Advancing sensory neuroprosthetics using artificial brain networks. *Patterns (N Y), 2*(7), 100304. https://doi.org/10.1016/j.patter.2021.100304.

Haufe, S., Meinecke, F., Görgen, K., Dähne, S., Haynes, J.-D., Blankertz, B., & Bießmann, F. (2014). On the interpretation of weight vectors of linear models in multivariate neuroimaging. *Neuroimage, 87*, 96–110.

He, Y., Eguren, D., Azorin, J. M., Grossman, R. G., Luu, T. P., & Contreras-Vidal, J. L. (2018). Brain-machine interfaces for controlling lower-limb powered robotic systems. *Journal of Neural Engineering, 15*(2), 021004. https://doi.org/10.1088/1741-2552/aaa8c0.

Heaton, J. B., Polson, N. G., & Witte, J. H. (2017). Deep learning for finance: Deep portfolios. *Applied Stochastic Models in Business and Industry, 33*(1), 3–12.

Hochberg, L. R., Bacher, D., Jarosiewicz, B., Masse, N. Y., Simeral, J. D., Vogel, J., ... Van Der Smagt, P. (2012). Reach and grasp by people with tetraplegia using a neurally controlled robotic arm. *Nature, 485*(7398), 372–375.

Holzinger, A., Langs, G., Denk, H., Zatloukal, K., & Müller, H. (2019). Causability and explainability of artificial intelligence in medicine. *Wiley Interdisciplinary Reviews: Data Mining and Knowledge Discovery, 9*(4), e1312.

Iberite, F., Muheim, J., Akouissi, O., Gallo, S., Rognini, G., Morosato, F., ... Shokur, S. (2023). Restoration of natural thermal sensation in upper-limb amputees. *Science (New York, N. Y.), 380*(6646), 731–735. https://doi.org/10.1126/science.adf6121.

Ienca, M., & Ignatiadis, K. (2020). Artificial intelligence in clinical neuroscience: Methodological and ethical challenges. *Ajob Neuroscience, 11*(2), 77–87.

Ienca, M., Kressig, R. W., Jotterand, F., & Elger, B. (2017). Proactive ethical Design for Neuroengineering, assistive and rehabilitation technologies: The Cybathlon lesson. *Journal of Neuroengineering and Rehabilitation, 14*(1), 1–11.

Iivanainen, J., Stenroos, M., & Parkkonen, L. (2017). Measuring MEG closer to the brain: Performance of on-scalp sensor arrays. *Neuroimage, 147*, 542–553. https://doi.org/10.1016/j.neuroimage.2016.12.048.

Jobin, A., Ienca, M., & Vayena, E. (2019). The global landscape of AI ethics guidelines. *Nature Machine Intelligence, 1*(9), 389–399.

Jotterand, F., & Ienca, M. (2021). *Artificial intelligence in brain and mental health: Philosophical, ethical & policy issues*. Springer.

Kamiya, J. (1962). *Conditioned discrimination of the EEG alpha rhythm in humans*. San Francisco, CA: Western Psychological Association.

Koelewijn, A. D., Audu, M., del-Ama, A. J., Colucci, A., Font-Llagunes, J. M., Gogeascoechea, A., ... Beckerle, P. (2021). Adaptation strategies for personalized gait neuroprosthetics [perspective]. *Frontiers in Neurorobotics, 15*(170), https://doi.org/10.3389/fnbot.2021.750519.

Kohl, S. H., Melies, P., Uttecht, J., Lührs, M., Bell, L., Mehler, D. M. A., ... Konrad, K. (2023). Successful modulation of temporoparietal junction activity and stimulus-driven attention by fNIRS-based neurofeedback—A randomized controlled proof-of-concept study. *Imaging Neuroscience, 1*, 1–26. https://doi.org/10.1162/imag_a_00014.

Kornhuber, H. H., & Deecke, L. (1964). Hirnpotentialänderungen beim Menschen vor und nach Willkürbewegungen, dargestellt mit Magnetbandspeicherung und Rückwärtsanalyse. *Pflügers Archiv, 281*(1), 52.

Krizhevsky, A., Sutskever, I., & Hinton, G. (2012). ImageNet classification with deep convolutional neural networks. *Neural Information Processing Systems, 25*. https://doi.org/10.1145/3065386.

Krol, L. R., Andreessen, L. M., & Zander, T. O. (2018). *Passive brain–computer interfaces: A perspective on increased interactivity. Brain–computer interfaces handbook: Technological and theoretical advances*. CRC Press, 69–86.

Krol, L. R., Haselager, P., & Zander, T. O. (2020). Cognitive and affective probing: A tutorial and review of active learning for neuroadaptive technology. *Journal of Neural Engineering, 17*(1), 012001. https://doi.org/10.1088/1741-2552/ab5bb5.

Kwon, O. Y., Lee, M. H., Guan, C., & Lee, S. W. (2020). Subject-independent brain–computer interfaces based on deep convolutional neural networks. *IEEE Transactions on Neural Networks and Learning Systems, 31*(10), 3839–3852. https://doi.org/10.1109/tnnls.2019.2946869.

Leta Jones, M., & Millar, J. (2017). *Hacking metaphors in the anticipatory governance of emerging technology: The case of regulating robots. The Oxford handbook of law, regulation and technology.* Oxford University Press, 597–619. https://doi.org/10.1093/oxfordhb/9780199680832.013.34.

Leuthardt, E. C., Moran, D. W., & Mullen, T. R. (2021). Defining surgical terminology and risk for brain computer interface technologies [perspective]. *Frontiers in Neuroscience, 15.* https://doi.org/10.3389/fnins.2021.599549.

Liao, S. M. (2023). Ethics of AI and health care: Towards a substantive human rights framework. *Topoi,* 1–10.

Liew, S. L., Rana, M., Cornelsen, S., Fortunato de Barros Filho, M., Birbaumer, N., Sitaram, R., ... Soekadar, S. R. (2016). Improving motor corticothalamic communication after stroke using real-time fMRI connectivity-based neurofeedback. *Neurorehabilitation and Neural Repair, 30*(7), 671–675. https://doi.org/10.1177/1545968315619699.

Lotte, F., Bougrain, L., Cichocki, A., Clerc, M., Congedo, M., Rakotomamonjy, A., & Yger, F. (2018). A review of classification algorithms for EEG-based brain–computer interfaces: A 10 year update. *Journal of Neural Engineering, 15*(3), 031005. https://doi.org/10.1088/1741-2552/aab2f2.

Makin, J. G., Moses, D. A., & Chang, E. F. (2020). Machine translation of cortical activity to text with an encoder-decoder framework. *Nature Neuroscience, 23*(4), 575–582. https://doi.org/10.1038/s41593-020-0608-8.

Metzger, S. L., Littlejohn, K. T., Silva, A. B., Moses, D. A., Seaton, M. P., Wang, R., ... Chang, E. F. (2023). A high-performance neuroprosthesis for speech decoding and avatar control. *Nature, 620*(7976), 1037–1046. https://doi.org/10.1038/s41586-023-06443-4.

Mitchell, M. (2023). How do we know how smart AI systems are? *Science (New York, N. Y.), 381*(6654), adj5957. https://doi.org/10.1126/science.adj5957.

Mnih, V., Kavukcuoglu, K., Silver, D., Graves, A., Antonoglou, I., Wierstra, D., & Riedmiller, M. (2013). Playing atari with deep reinforcement learning. *arXiv preprint arXiv:1312.5602.*

Moses, D. A., Leonard, M. K., Makin, J. G., & Chang, E. F. (2019). Real-time decoding of question-and-answer speech dialogue using human cortical activity. *Nature Communications, 10*(1), 3096. https://doi.org/10.1038/s41467-019-10994-4.

Muller-Putz, G. R., Scherer, R., Brauneis, C., & Pfurtscheller, G. (2005). Steady-state visual evoked potential (SSVEP)-based communication: Impact of harmonic frequency components. *Journal of Neural Engineering, 2*(4), 123–130. https://doi.org/10.1088/1741-2560/2/4/008.

Muthukumaraswamy, S. (2013). High-frequency brain activity and muscle artifacts in MEG/EEG: A review and recommendations [Review]. *Frontiers in Human Neuroscience, 7.* https://doi.org/10.3389/fnhum.2013.00138.

Nann, M., Cohen, L. G., Deecke, L., & Soekadar, S. R. (2019). To jump or not to jump—The Bereitschaftspotential required to jump into 192-meter abyss. *Scientific Reports, 9*(1), 2243. https://doi.org/10.1038/s41598-018-38447-w.

Nasr, K., Haslacher, D., Dayan, E., Censor, N., Cohen, L. G., & Soekadar, S. R. (2022). Breaking the boundaries of interacting with the human brain using adaptive closed-loop stimulation. *Progress in Neurobiology, 216*, 102311. https://doi.org/10.1016/j.pneurobio.2022.102311.

Nasr, K., Haslacher, D., & Soekadar, S. (2023). *Towards adaptive deep brain neuromodulation using temporal interference magnetic stimulation, 10.* Transcranial Magnetic Stimulation (TMS),.

Obermeyer, Z., Powers, B., Vogeli, C., & Mullainathan, S. (2019). Dissecting racial bias in an algorithm used to manage the health of populations. *Science (New York, N. Y.), 366*(6464), 447–453. https://doi.org/10.1126/science.aax2342.

O'Doherty, J. E., Lebedev, M. A., Ifft, P. J., Zhuang, K. Z., Shokur, S., Bleuler, H., & Nicolelis, M. A. (2011). Active tactile exploration using a brain-machine-brain interface. *Nature, 479*(7372), 228–231. https://doi.org/10.1038/nature10489.

Owolabi, M. O., Leonardi, M., Bassetti, C., Jaarsma, J., Hawrot, T., Makanjuola, A. I., ... Servadei, F. (2023). Global synergistic actions to improve brain health for human development. *Nature Reviews Neurology, 19*(6), 371–383. https://doi.org/10.1038/s41582-023-00808-z.

Oxley, T. J., Yoo, P. E., Rind, G. S., Ronayne, S. M., Lee, C. M. S., Bird, C., ... Opie, N. L. (2021). Motor neuroprosthesis implanted with neurointerventional surgery improves capacity for activities of daily living tasks in severe paralysis: First in-human experience. *Journal of NeuroInterventional Surgery, 13*(2), 102–108. https://doi.org/10.1136/neurintsurg-2020-016862.

Penfield, W., & Boldrey, E. (1937). Somatic motor and sensory representation in the cerebral cortex of man as studied by electrical stimulation. *Brain, 60*(4), 389–443.

Peterson, F., & Jung, C. G. (1907). Psycho-physical investigations with the galvanometer and pneumograph in normal and insane individuals. *Brain, 30*(2), 153–218. https://doi.org/10.1093/brain/30.2.153.

Ramos-Murguialday, A., Broetz, D., Rea, M., Läer, L., Yilmaz, O., Brasil, F. L., ... Birbaumer, N. (2013). Brain-machine interface in chronic stroke rehabilitation: A controlled study. *Annals of Neurology, 74*(1), 100–108. https://doi.org/10.1002/ana.23879.

Richards, N., & Smart, W. (2013). How should the law think about robots? *SSRN Electronic Journal.* https://doi.org/10.2139/ssrn.2263363.

Romeni, S., Zoccolan, D., & Micera, S. (2021). A machine learning framework to optimize optic nerve electrical stimulation for vision restoration. *Patterns* (this issue).

Rumelhart, D. E., Hinton, G. E., & Williams, R. J. (1986). Learning representations by back-propagating errors. *Nature, 323*(6088), 533–536.

Sand, M., Durán, J. M., & Jongsma, K. R. (2022). Responsibility beyond design: Physicians' requirements for ethical medical AI. *Bioethics, 36*(2), 162–169.

Scangos, K. W., Khambhati, A. N., Daly, P. M., Makhoul, G. S., Sugrue, L. P., Zamanian, H., ... Chang, E. F. (2021). Closed-loop neuromodulation in an individual with treatment-resistant depression. *Nature Medicine, 27*(10), 1696–1700. https://doi.org/10.1038/s41591-021-01480-w.

Scangos, K. W., Makhoul, G. S., Sugrue, L. P., Chang, E. F., & Krystal, A. D. (2021). State-dependent responses to intracranial brain stimulation in a patient with depression. *Nature Medicine, 27*(2), 229–231. https://doi.org/10.1038/s41591-020-01175-8.

Sellers, E. W., & Donchin, E. (2006). A P300-based brain-computer interface: Initial tests by ALS patients. *Clinical Neurophysiology: Official Journal of the International Federation of Clinical Neurophysiology, 117*(3), 538–548. https://doi.org/10.1016/j.clinph.2005.06.027.

Simon, C., Bolton, D. A. E., Kennedy, N. C., Soekadar, S. R., & Ruddy, K. L. (2021). Challenges and opportunities for the future of brain-computer interface in neurorehabilitation. *Front Neurosci, 15*, 699428. https://doi.org/10.3389/fnins.2021.699428.

Sitaram, R., Ros, T., Stoeckel, L., Haller, S., Scharnowski, F., Lewis-Peacock, J., ... Sulzer, J. (2017). Closed-loop brain training: The science of neurofeedback. *Nature Reviews. Neuroscience, 18*(2), 86–100. https://doi.org/10.1038/nrn.2016.164.

Sitaram, R., Weiskopf, N., Caria, A., Veit, R., Erb, M., & Birbaumer, N. (2007). fMRI brain-computer interfaces. *IEEE Signal processing Magazine, 25*(1), 95–106.

Soekadar, S. R., Born, J., Birbaumer, N., Bensch, M., Halder, S., Murguialday, A. R., ... Martens, S. (2013). Fragmentation of slow wave sleep after onset of complete locked-in state. *Journal of Clinical Sleep Medicine: JCSM: Official Publication of the American Academy of Sleep Medicine, 9*(9), 951–953. https://doi.org/10.5664/jcsm.3002.

Soekadar, S. R., Kohl, S. H., Mihara, M., & von Lühmann, A. (2021). Optical brain imaging and its application to neurofeedback. *NeuroImage: Clinical, 30*, 102577. https://doi.org/10.1016/j.nicl.2021.102577.

Soekadar, S.R., & Nasr, K. (2023). *System for stimulation of biological tissue* (EP Patent No).

Soekadar, S. R., Vermehren, M., Colucci, A., Haslacher, D., Bublitz, C., Ienca, M., ... Blankertz, B. (2023). Future developments in brain/neural–computer interface technology. In V. Dubljević, & A. Coin (Eds.). *Policy, identity, and neurotechnology* (pp. 65–85) Springer International Publishing. https://link.springer.com/10.1007/978-3-031-26801-4_5.

Soekadar, S. R., Witkowski, M., Birbaumer, N., & Cohen, L. G. (2015). Enhancing Hebbian learning to control brain oscillatory activity. *Cerebral Cortex, 25*(9), 2409–2415. https://doi.org/10.1093/cercor/bhu043.

Soekadar, S. R., Witkowski, M., Gómez, C., Opisso, E., Medina, J., Cortese, M., ... Vitiello, N. (2016). Hybrid EEG/EOG-based brain/neural hand exoskeleton restores fully independent daily living activities after quadriplegia. *Science Robotics, 1*(1), eaag3296. https://doi.org/10.1126/scirobotics.aag3296 http://robotics.sciencemag.org/content/1/1/eaag3296.abstract.

Soekadar, S. R., Witkowski, M., Vitiello, N., & Birbaumer, N. (2015). An EEG/EOG-based hybrid brain-neural computer interaction (BNCI) system to control an exoskeleton for the paralyzed hand. *Biomedizinische Technik (Berlin), 60*(3), 199–205. https://doi.org/10.1515/bmt-2014-0126.

Starke, G., De Clercq, E., Borgwardt, S., & Elger, B. S. (2021). Computing schizophrenia: Ethical challenges for machine learning in psychiatry. *Psychological Medicine, 51*(15), 2515–2521.

Starke, G., De Clercq, E., & Elger, B. S. (2021). Towards a pragmatist dealing with algorithmic bias in medical machine learning. *Medicine, Health Care, and Philosophy, 24*, 341–349.

Stawicki, P., Gembler, F., & Volosyak, I. (2016). Driving a semiautonomous mobile robotic car controlled by an SSVEP-based BCI. *Computational Intelligence and Neuroscience, 2016*, 4909685. https://doi.org/10.1155/2016/4909685.

Steinert, S., Bublitz, C., Jox, R., & Friedrich, O. (2019). Doing things with thoughts: Brain-computer interfaces and disembodied agency. *Philosophy & Technology, 32*(3), 457–482.

Sterman, M. B., Wyrwicka, W., & Howe, R. (1969). Behavioral and neurophysiological studies of the sensorimotor rhythm in the cat. *Electroencephalography and Clinical Neurophysiology, 27*(7), 678–679. https://doi.org/10.1016/0013-4694(69)91281-4.

Sünderhauf, N., Brock, O., Scheirer, W., Hadsell, R., Fox, D., Leitner, J., ... Milford, M. (2018). The limits and potentials of deep learning for robotics. *The International Journal of Robotics Research, 37*(4–5), 405–420.

Swann, N. C., de Hemptinne, C., Miocinovic, S., Qasim, S., Ostrem, J. L., Galifianakis, N. B., ... Starr, P. A. (2018). Chronic multisite brain recordings from a totally implantable bidirectional neural interface: Experience in 5 patients with Parkinson's disease. *Journal of Neurosurgery, 128*(2), 605–616. https://doi.org/10.3171/2016.11.Jns161162.

Tang, J., LeBel, A., Jain, S., & Huth, A. G. (2023). Semantic reconstruction of continuous language from non-invasive brain recordings. *Nature Neuroscience, 26*(5), 858–866. https://doi.org/10.1038/s41593-023-01304-9.

Thielen, J., Marsman, P., Farquhar, J., & Desain, P. (2021). From full calibration to zero training for a code-modulated visual evoked potentials for brain–computer interface. *Journal of Neural Engineering, 18*(5), https://doi.org/10.1088/1741-2552/abecef.

Ushiba, J., & Soekadar, S. R. (2016). Brain-machine interfaces for rehabilitation of post-stroke hemiplegia. *Progress in Brain Research, 228*, 163–183. https://doi.org/10.1016/bs.pbr.2016.04.020.

Vayena, E., Blasimme, A., & Cohen, I. G. (2018). Machine learning in medicine: Addressing ethical challenges. *PLoS Medicine, 15*(11), e1002689.

Venthur, B., Blankertz, B., Gugler, M. F., & Curio, G. (2010, 10–13 Oct. 2010). Novel applications of BCI technology: Psychophysiological optimization of working conditions in industry. In *2010 IEEE international conference on systems, man and cybernetics*.

Vidal, J. J. (1973). Toward direct brain-computer communication. *Annual Review of Biophysics and Bioengineering, 2*, 157–180. https://doi.org/10.1146/annurev.bb.02.060173.001105.

Walter, W. G. (1943). An automatic low frequency analyser. *Electronic Engineering, 16*, 9–13.

Walter, W. G., Cooper, R., Aldridge, V. J., McCallum, W. C., & Winter, A. L. (1964). Contingent negative variation: An electric sign of sensori-motor association and expectancy in the human brain. *Nature, 203*(4943), 380–384. https://doi.org/10.1038/203380a0.

Wang, M., Li, R. J., Zhang, R. F., Li, G. Y., & Zhang, D. G. (2018). A wearable SSVEP-based BCI system for quadcopter control using head-mounted device. *IEEE Access, 6*. https://doi.org/10.1109/Access.2018.2825378.

Weissglass, D. E. (2022). Contextual bias, the democratization of healthcare, and medical artificial intelligence in low-and middle-income countries. *Bioethics, 36*(2), 201–209.

Willett, F. R., Avansino, D. T., Hochberg, L. R., Henderson, J. M., & Shenoy, K. V. (2021). High-performance brain-to-text communication via handwriting. *Nature, 593*(7858), 249–254. https://doi.org/10.1038/s41586-021-03506-2.

Willett, F. R., Kunz, E. M., Fan, C., Avansino, D. T., Wilson, G. H., Choi, E. Y., ... Henderson, J. M. (2023). A high-performance speech neuroprosthesis. *Nature, 620*(7976), 1031–1036. https://doi.org/10.1038/s41586-023-06377-x.

Wolpaw, J. R. (2013). Chapter 6—Brain–computer interfaces. In M. P. Barnes, & D. C. Good (Vol. Eds.), *Handbook of clinical neurology: Vol. 110*, (pp. 67–74). (pp. 67) Elsevier. https://doi.org/10.1016/B978-0-444-52901-5.00006-X.

Zander, T. O., & Jatzev, S. (2012). Context-aware brain-computer interfaces: Exploring the information space of user, technical system and environment. *Journal of Neural Engineering, 9*(1), 016003. https://doi.org/10.1088/1741-2560/9/1/016003.

Zander, T. O., & Kothe, C. (2011). Towards passive brain-computer interfaces: Applying brain-computer interface technology to human-machine systems in general. *Journal of Neural Engineering, 8*(2), 025005. https://doi.org/10.1088/1741-2560/8/2/025005.

Zander, T. O., Krol, L. R., Birbaumer, N. P., & Gramann, K. (2016). Neuroadaptive technology enables implicit cursor control based on medial prefrontal cortex activity. *Proceedings of the National Academy of Sciences of the United States of America, 113*(52), 14898–14903. https://doi.org/10.1073/pnas.1605155114.

Zerfowski, J., Sander, T. H., Tangermann, M., Soekadar, S. R., & Middelmann, T. (2021). Real-time data processing for brain-computer interfacing using optically pumped magnetometers. *International Journal of Bioelectromagnetism, 23*(1), 14. http://www.ijbem.org/volume23/number2/14.pdf.

Zhang, Z., Geiger, J., Pohjalainen, J., Mousa, A. E.-D., Jin, W., & Schuller, B. (2018). Deep learning for environmentally robust speech recognition: An overview of recent developments. *ACM Transactions on Intelligent Systems and Technology (TIST), 9*(5), 1–28.

Zollo, L., Di Pino, G., Ciancio, A. L., Ranieri, F., Cordella, F., Gentile, C., ... Uglielmelli, E. (2019). Restoring tactile sensations via neural interfaces for real-time force-and-slippage closed-loop control of bionic hands. *Science Robotics, 4*(27), https://doi.org/10.1126/scirobotics.aau9924.

CHAPTER TWO

Computational psychiatry and AI – High hopes: heralded heights or hollow hype?

Derya Şahin[*]
Stiftung Mercator, Essen, Germany
*Corresponding author. e-mail address: deryasahin@protonmail.ch

Contents

1. Introduction	29
2. General AI-related ethical concerns	31
2.1 Data privacy and security	31
2.2 Bias and fairness	31
2.3 Explainability	32
3. Ethical concerns specific to computational psychiatry	32
3.1 Hypothesis-driven versus data-driven approaches	32
3.2 Reductionism	34
3.3 Understanding versus knowing	37
3.4 Constraints through the quality and nature of data in psychiatry	39
4. Conclusion	42
References	42

Abstract

Computational psychiatry is a multidisciplinary field that utilizes mathematical, statistical, and computational methods to better understand mental disorders. The integration of AI in computational psychiatry has opened new possibilities for creating more precise and nuanced models of psychiatric disorders, simultaneously raising important ethical concerns related to privacy, data security, transparency, bias, alignment, and limits of computational psychiatry. This chapter provides an overview of the ethical considerations and challenges of computational psychiatry and the use of AI, specifically related to nosology, reductionism, and data constraints specific to psychiatry. Finally, it questions the epistemological limits of computational psychiatry.

1. Introduction

Computational psychiatry is a multidisciplinary field that utilizes mathematical, statistical, and computational methods to better understand

---

Developments in Neuroethics and Bioethics, Volume 7
ISSN 2589-2959, https://doi.org/10.1016/bs.dnb.2024.02.013
Copyright © 2024 Elsevier Inc. All rights are reserved, including those for text and data mining, AI training, and similar technologies.

and treat mental disorders (Montague, Dolan, Friston, & Dayan, 2012). Its applications comprise a wide spectrum of purposes, from understanding the neural underpinnings of psychiatric diseases to developing personalized treatments for them. The progress in artificial intelligence (AI) is affecting computational psychiatry by both offering new methods of research and new methods of implementation.

Computational psychiatry is a data-driven approach for which AI offers unprecedented tools with its capability of analyzing enormous amounts of multimodal data (Rutledge, Chekroud, & Huys, 2019). AI offers an advantage in contrast to traditional data analysis methods especially regarding pattern recognition and predictive analysis (Bennett, Hayes, Kleczyk, & Mehta, 2022). In psychiatry, predictive modelling that is made viable by AI can be a useful tool for estimating risk of developing certain mental health conditions and improving early intervention as well as personalizing prevention strategies. The diagnosis of mental health disorders can be improved by employing methods such as analyses of speech, facial expressions, data collected by wearables, digital behavioural data (collected by social media or smartphone use) (Birnbaum et al., 2019; Ríssola, Losada, & Crestani, 2021; Thorstad & Wolff, 2019). Also, the analysis of the changes in the patterns of behaviour can be monitored better by incorporating AI into computational psychiatry research (Parziale & Mascalzoni, 2022).

A potential conflict arises between the epistemic and ethical goals of psychiatric research and practice when applying computational methods to them. Epistemic goals refer here to extracting the maximum possible amount of information from data and achieving high diagnostic and predictive accuracy; whereas ethical goals refer to not causing harm (non-maleficence), being useful (beneficence), ensuring autonomy and justice (Beauchamp & Childress, 2013). The first report of the World Health Organisation on Ethics and Governance of Artificial Intelligence for Health (Health Ethics & Governance, 2021) from June 2021 emphasises that researchers and clinicians should adopt an "ethics by design" approach for AI tools. Ethics by design refers to the integration and continuous examination of ethical principles in AI tools.

The ethical concerns regarding AI-supported computational psychiatry are in addition to typical AI-related issues such as privacy and data security, bias, fairness, and explainability, as well as more specific issues related to the use of AI in computational psychiatry, which are related to reductionism, the difference between understanding vs. knowing, stigmatisation, nosology as well as data constraints specific to psychiatry.

2. General AI-related ethical concerns
2.1 Data privacy and security

Data collected in psychiatry contain information about perception, feelings, thoughts and behaviour of human beings - highly personal and sensitive information. The protection of such data is essential, since the risks of misuse are paramount. Third-party access or data breaches to psychiatric data can facilitate malicious use, such as discrimination, blackmailing, identity theft or fraud. Although data are stored in an anonymised form, the possibility of de-anonymising data still exists (Guru Prasad, Gujjar, Naveen Kumar, Anand Kumar, & Chandrappa, 2023; Howe, Iii, & Elenberg, 2020) and with the increasing sophistication of cyberattacks, establishing security measures that guarantee protection against cyber threats and unauthorised access is becoming increasingly difficult (Penfold, 2023).

2.2 Bias and fairness

Bias and fairness are core ethical concerns in AI. Fairness refers to the pursuit of unbiased algorithms and is strongly associated with the data used for developing models. Although AI models based on large datasets excel at recognizing complex patterns and making highly accurate predictions, the algorithms reflect the characteristics of the datasets they were trained on (Cowgill et al., 2020). When trained with biased data, AI as a data-driven approach will carry forward the biases incorporated in the data (Suresh & Guttag, 2021). Unfortunately, data collected and used for medical research often contain bias—especially representation bias—due to a number of reasons, to which (computational) psychiatry is no exemption. First, most of the biomedical research is conducted in high-income regions, leading to an underrepresentation of low- and middle-income regions comprising two thirds of the world population (Sumathipala, Siribaddana, & Patel, 2004). Second, the data collected for biomedical research in Europe and North America usually overrepresent well-educated, socioeconomically advantaged, white people (Baquet, Commiskey, Daniel Mullins, & Mishra, 2006; Garber & Arnold, 2006). Third, the sensitive nature of medical data impedes data sharing, preventing the analysis of merged data from multiple sources.

Previous research hints that algorithms trained on biological features might be substantially less accurate for people of colour (Sahin, Jessen, & Kambeitz, 2022), psychosis prediction might result in biases depending on educational status (Şahin et al., 2023), and recidivism algorithms might show racial bias (Dressel & Farid, 2018). Ensuring equitable treatment for all individuals influenced by algorithms, particularly historically marginalised groups such as women, ethnic/racial minorities, and those with lower socioeconomic status, is imperative. It is essential to proactively prevent the exacerbation of disparities through the inadvertent biases embedded within these algorithms.

2.3 Explainability

Explainability (also referred to as interpretability or transparency) defines the degree to which the decisions or outcomes of an algorithm are transparent or intelligible (Floridi et al., 2018). Black-box algorithms—non-interpretable algorithms that cannot be deciphered to understand how they infer their outcomes—are challenging for computational psychiatry from a number of aspects: they can render computational psychiatry more mechanical and reductionist (discussed in detail below in sections *Reductionism* and *Understanding* vs. *Knowing*) and are highly relevant for informed consent: the nature and dimensions of information that can be extracted from or inferred through AI using psychiatric data are not necessarily known to the full extent before the development of an algorithm (Christen, Domingo-Ferrer, Draganski, Spranger, & Walter, 2016). For example, a study concluded that AI could predict the race of an individual using imaging data, while the mechanism underlying this prediction remained elusive to the researchers, lacking an explicable rationale (Gichoya et al., 2022). If the source of the data (in this case, the subject whose data are collected for computational psychiatry) lacks awareness regarding the potential inferences derivable from their data, the discernment of their consent's informed nature is cast into doubt (Christen et al., 2016).

3. Ethical concerns specific to computational psychiatry

3.1 Hypothesis-driven versus data-driven approaches

Perception without conception is blind; conception without perception is empty.
Immanuel Kant, Critique of Pure Reason, Kant (1787) (AA 03: 75)

In order to be able to understand the psychiatry-specific ethical issues that arise from the use of computational methods in psychiatry, one should first illuminate the approaches to knowledge generation in computational psychiatry, namely hypothesis-driven vs. data-driven approaches, as the ethical issues illuminated in this chapter mainly arise from the latter. Karl Friston differentiates between hypothesis- and data-driven approaches as strong and weak approaches (Friston, 2023): strong approaches are hypothesis- or theory-driven approaches that form generative models, whereas weak approaches are data-driven, theory-free, and descriptive.

Data-driven approaches in computational psychiatry involve analyzing extensive datasets without predefined hypotheses, aiming to extract patterns, relationships, and insights directly from the data. Advanced techniques such as machine learning are utilized to discern meaningful correlations and trends. On the one hand, this approach is advantageous in that it can unveil unexpected and previously unnoticed associations within the data, making it especially useful for exploratory analysis. On the other hand, it might lead to spurious correlations, misinterpretations, and a lack of robust understanding (Coveney, Dougherty, & Highfield, 2016; Wiese & Friston, 2022).

Hypothesis-driven approaches start with a specific hypothesis or theoretical framework based on existing knowledge or assumptions. Data are then collected and analysed to either validate or refute these predetermined hypotheses. Thus, a clear theoretical foundation is established for the research, allowing for targeted testing of specific ideas. However, hypothesis-driven approaches can be limiting as they might overlook unexpected patterns or nuances that do not align with the initial hypothesis and could potentially miss out on novel discoveries.

In a 2008 article, Chris Anderson argued that the increasing volume of data and the computational methods in the digital age would make the traditional scientific method of forming hypotheses and testing them unnecessary (Anderson, 2008). Instead, he suggested that with vast amounts of data, correlations could be identified without needing to understand the underlying mechanisms or causation.

However, while data-driven approaches are powerful, they are not a substitute for theoretical understanding. Causation, context, and the need for explanatory models are crucial elements in scientific inquiry and increasing amounts of data do not always necessarily result in increasing amounts of knowledge (Coveney et al., 2016; Succi & Coveney, 2019). Especially, as data-driven approaches are based on error minimisation, their applicability to complex systems such as human brain and behaviour is doomed to be limited (Succi & Coveney, 2019).

In practice, while a combination of both approaches is often found to be the most effective, it is not always possible. Data-driven techniques can reveal patterns that generate new hypotheses, which can then be tested and validated using a hypothesis-driven approach, thus providing a more comprehensive understanding in the realm of computational psychiatry. Without translating findings of data-driven methods to hypotheses and uncovering clear causal pathways, improving psychiatric practice would be challenging. To date, computational psychiatry has not been successful in delivering causal explanations or improving our understanding of psychiatric diseases in a manner that connects molecular and neurobiological levels of the human brain with more complex and macro level phenomena such as feelings, thoughts, behaviour, perception and consciousness (Huys, Maia, & Frank, 2016).

3.2 Reductionism

What would you call this patient—schizophrenic or schizoaffective?
I think I'd call him Michael McIntyre.
Bessel van der Kolk, The Body Keeps the Score, Van der Kolk (2014)

Reductionism in computational psychiatry refers to simplifying complex mental processes or phenomena into fundamental components, often at the neurobiological or computational level, to better understand and model psychiatric disorders (Adams, Huys, & Roiser, 2016; Kirmayer & Gold, 2011). This approach seeks to explain higher-level mental functions by breaking them down into smaller, more manageable parts. It assumes that understanding the basic components and their interactions will lead to a comprehensive understanding of mental disorders and their underlying mechanisms.

Psychiatric phenomena, however, exhibit emergent properties, being characterized by a greater sum of effects than its individual elements. The brain, a remarkably intricate and specialised organ, epitomises this complexity—each neuron functioning uniquely, posing significant challenges in encoding and modelling brain function. The intricate interplay of genes and life experiences, starting even within the womb, shapes and influences our brain's wiring, resulting in the high degree of individuality and complexity observed in every human being. Thus, psychiatric disorders embody the distinctive interplay of an individual's inherent characteristics and their upbringing, influencing their learning, thought processes, emotions, reactions, interactions, and coping mechanisms. The aetiology of psychiatric diseases is expected to be highly individual in each patient's case, almost paralleling the opening lines of Tolstoy's *Anna Karenina*: *"Happy families are all alike; every unhappy family is unhappy in its own way"*. Although computational approaches can offer valuable insights, relying exclusively on computational methods might overlook the emotional and psychological dimensions crucial to effective mental health treatment. Considering that psychiatric diagnoses are arbitrary constructs without a homogeneous and specific aetiology (Kendler, 2016; Zachar & Kendler, 2017), an exclusive focus on the brain in computational psychiatry could result in overlooking critical aspects. Such a reductionist perspective would not adequately address the intricate nature of mental health disorders and the various factors influencing their progression (Wiese & Friston, 2022).

To exemplify, one can take a closer look at depression as a psychiatric condition. Depression is a syndrome that manifests through low mood, lack of interest, listlessness, inability to feel pleasure, sleep troubles, to name a few symptoms (Malhi & Mann, 2018). While the diagnosis of the disease is based on the presence and the number of symptoms that are typical for a depressive syndrome, the exact manifestation in each individual is different, and the attributed causes of the syndrome can vary from psychosocial burden to metabolic dysregulations to brain diseases (tumours, Parkinson's disease, Alzheimer's disease) to nutritional deficiencies and comprise dozens of different potential roots. Furthermore, past experiences, thinking and belief patterns, life conditions, social environment, genetics, and many other factors can facilitate the outbreak of the condition or prevent it. Research so far has not shown a specific area, pathway or process in the brain that underlies all depressions or whose disturbance leads to a clinical depression in every individual (Chaudhury, Liu, & Han, 2015). Attempts at capturing a syndrome so heterogeneous by computational methods and explaining it as a disturbance in some neural circuits would inevitably result in the simplification of the condition.

Still, it cannot be completely ruled out that in the future the complex and highly individual nature of psychiatric conditions can be depicted and understood as computational models and data points. However, apart from the challenge that arises from the complex and emergent nature of psychiatric disease, a computational understanding of the human brain and behaviour can result in transformative effects (Wiese & Friston, 2022) that would not only force us to rethink the meaning of free will, autonomy and accountability, but also have substantial effects on the way patients perceive themselves and patients, practitioners and caregivers engage in treatments.

Both reductionism and the nature of approaches (hypothesis- or data-driven) would have implications for the treatments that would emerge from computational approaches to psychiatry. If the understanding of psychiatric conditions shifts to conditions caused by an imbalance of neurotransmitters or disturbances of neural circuits and brain connectivity, it could potentially mislead the patients as well as the practitioners to believe that the conditions can only be addressed by biological treatments (i.e., medication, brain stimulation, surgical intervention), reduce the understanding for the social and environmental conditions that are relevant to the emergence of the disease and increase the associated stigma (Angermeyer, Holzinger, Carta, & Schomerus, 2011; Lebowitz & Ahn, 2014). Thus, the importance of restructuring cognition, beliefs and behaviour as well as the significance of one's own actions and agency can be undermined, which can cause or reinforce overattachment to diagnoses in patients, leading to externalisation and taking passive roles toward their treatment (Paris, 2020).

Furthermore, following questions with potentially transformative effects arise from computational approaches to psychiatry: If we manage to fully decode the mechanisms that underlie human feelings, thoughts, perceptions and behaviour through computational models, what will the implications for our own understanding of being human be (Fuchs, 2006; Huys, Moutoussis, & Williams, 2011)? How would we define accountability and autonomy in a world in which the functioning of the human brain is fully deciphered? Would the ability to pinpoint the exact mechanisms creating our feelings, thoughts, perceptions, and behaviour make them less real or relevant? If my poor living conditions and high psychosocial burden cause depression, is it sufficient and, more importantly, ethically acceptable to simply *fix* the circuit in my brain that is eliciting the depression? Or manipulate our brains so that, no matter what happens in one's surroundings and to oneself, negative emotions (sadness, anger, shame) are not being elicited anymore? It might be important to conceive and approach the human condition—with psychiatric conditions as an undeniable part of it—not merely as a mechanical phenomenon that can be decomposed as signal processing between neuronal networks. It is equally important to question whether it is sufficient to eliminate disease, with a focus on optimising the functionality of human beings while overlooking the broader environmental and societal conditions in which human beings exist. By focusing on eliminating psychiatric diseases, one also neglects that some manifestations of psychiatric disease may represent adaptive reactions from an evolutionary standpoint.

3.3 Understanding versus knowing

One of the primary obstacles in ensuring the ethical application of AI, namely explainability, presents a simultaneous challenge within computational psychiatry. In general terms, as the sophistication and complexity of an algorithm's architecture increase, its operations become less comprehensible (Hassija et al., 2023; Huysmans, Baesens, & Vanthienen, 2006). For instance, while it's feasible to discern feature weights and deduce the contribution of each feature to the final decision in an algorithm based on a random forest, this is only partially achievable in a deep learning model. This initial challenge highlights the limited accessibility to information regarding how inferences are derived. Given that computational psychiatry models can encompass multimodal data, incorporating behavioural, genetic, imaging, and other biological features, and may encompass extensive data points per individual, it becomes imperative to discern which features significantly influenced the algorithmic decisions. Secondly, even when we manage to extract insights on feature importance in algorithmic decisions, it remains critical to differentiate between correlations and causality. As stated before, merely achieving accuracy within an algorithm does not ensure its validity, as true validity necessitates an understanding of the actual causal factors underlying a disorder.

The distinction between knowing and understanding an outcome holds particular significance in psychiatry compared to other medical fields. Psychiatric disorders, as syndrome-based constructs formed through consensus rather than uniform etiological foundations, pose a challenge for scientific disciplines like computational psychiatry. Relying on consensus-based constructs and corresponding data as the foundation for such a discipline compromises the reliability of its results. The data, being a second-order arbitrary construct reflective of consensus, introduces noise and raises doubts about the validity of the outcomes. Compounding this issue is the absence of objective external validation for psychiatric diagnoses due to the lack of homogeneous etiological diseases. Without such diseases, reliable biomarkers necessary for diagnosis and prediction remain elusive in the current state of psychiatry.

A predictive algorithm that accurately predicts a specific outcome for a patient but lacks explainability poses inherent dangers, particularly in psychiatry where the repercussions of an erroneous prediction can be severe. As an example, one can consider an algorithm designed to forecast the likelihood of an individual developing psychosis in their lifetime. A false positive prediction may lead to detrimental consequences such as societal stigma, unnecessary treatments with potential serious side effects, and psychological burden. In cases where the predicted outcome does not materialise, the price paid due to the false positive prediction could be substantial.

The crucial difference between understanding and knowing, coupled with the often-limited ability to establish causality within the field of psychiatry, can compound the difficulties of differentiating between biased judgments and sound predictions. If a particular trait is associated with a disease in a sample of patients, one could infer that the people carrying that trait are more prone to develop the associated disease. For example, education status is associated with the risk for developing psychosis, with people developing psychosis having a lower educational status compared to people who do not develop psychosis (Fusar-Poli et al., 2017). If one trains an algorithm based on data in which such an association is incorporated, the algorithm can learn to associate higher education with a lower risk of developing psychosis. Indeed, an analysis of psychosis prediction algorithms showed a general tendency to predict more favourable outcomes for individuals with higher educational level at baseline, which translated to a higher false positive rate in individuals with lower educational level (Şahin et al., 2023). Although there is an association between risk of developing psychosis and education years, there is no clear causal relation that has been shown between education and psychosis risk so far, and the nature of the association is likely to be highly complex and multifactorial, thus a clear one-directional causality is not to be expected. One could try to mitigate the bias in the algorithm by shifting the decision threshold (which would be diagnostic thresholds as soon as the algorithm is clinically deployed) in the one or the other direction, which would change the frequency of false positive and negative rates for the two groups (people with higher vs. lower educational level). The validity of the algorithm, however, cannot be improved by shifting decision thresholds. Increasing the validity of the algorithm would require understanding of what causes psychosis rather than what is associated with psychosis.

A study focusing on the proneness of psychiatrists to false advice from AI delivered interesting data on the clinical choices. The accuracy of psychiatrists' first treatment choices showed an approximately 35% (thus, not particularly brilliant) overlap with the expert consensus collected for the study and dropped further (to a bare 25%) when the algorithm gave an incorrect recommendation on which antidepressant to choose for the treatment (Jacobs et al., 2021). Before one discusses why the psychiatrists were susceptible to incorrect recommendations, one should first question the poor initial accuracy of treatment choices. Does this stem from the inadequacy of the participating psychiatrists' skills, or are there other factors contributing to the suboptimal accuracy? A probable explanation of the observed low accuracy of treatment choices is the well-recognized limitations of psychiatric classification systems (Barron, n.d.; Kendler, 2016; Zachar & Kendler, 2017). These constraints make it challenging to establish robust evidence on the most effective treatments for specific conditions, consequently leading to a lack of clear guidelines on how to approach psychiatric disease treatment.

3.4 Constraints through the quality and nature of data in psychiatry

While it is tempting to apply computational methods and AI in general to psychiatric research, one must carefully consider crucial constraints regarding the nature and quality of data that are still not sufficiently dealt with in psychiatry. Psychiatric diseases affect feelings, thoughts, behavior, and perception of individuals; consequently, studying psychiatric disease requires collecting data on these subjective elements, a task which is extremely challenging if not outright impossible to achieve through purely objective measurement methods.

There are several factors complicating the data collection on psychiatric disease and phenomena. First, self-report and observation as methods that are commonly used for data collection are highly variable both intra- and interpersonally, and are subject to subjectivity by their nature (Fuchs, 2010; Nordgaard, Sass, & Parnas, 2013). Second, full transparency of a person cannot be guaranteed all the time and on every matter, phenomena such as shame, guilt or social desirability may affect what one might disclose (Chan, 2010; Robins, Chris Fraley, & Krueger, 2009). Third, data collection often involves proxies rather than direct measures of what one is aiming to capture (Uusitalo, Tuominen, & Arstila, 2021). Thus, data on feelings, thoughts and perception of an individual is flawed at its best. To exemplify, one can look at questionnaires designed for assessing the severity of psychiatric diseases. The Beck Depression Inventory (Beck, Ward, Mendelson, Mock, & Erbauch, n.d.) is one such example that is a self-report instrument for assessing depression. In this questionnaire, the

Table 1 Beck depression inventory.

Example Item Nr. 1	Example Item Nr. 2
I don't feel disappointed by myself	I don't feel that I look any worse than I used to
I am disappointed in myself	I am worried that I am looking old or unattractive.
I am disgusted with myself	I feel there are permanent changes in my appearance that make me feel look unattractive.
I hate myself	I believe that I look ugly.

individual is instructed to choose from four statements the statement that best describes their state within the last week for 21 different items. Once one takes a closer look at the following two items from this questionnaire (See Table 1), the flawed nature of data that might be generated using the questionnaire becomes evident.

Subjects are instructed to choose the statement that best describes their state during the past week.

These two items as part of an instrument supposed to measure depression symptoms showcase that the data from such instruments are not robust, direct, and specific measures of a psychiatric condition, but rather proxies, highly subjective, non-specific, and thus, a measure of everything and nothing. What if someone is *ugly* by common standards? What if someone really has gone through permanent changes in their appearance? What if someone feels disappointed in and disgusted with themselves at the same time and hates themselves on top? What if someone feels disappointed in themselves because they did something that violated their morals? What if someone has felt disgusted with themselves for such a long time that it is rather a character trait of them than a sign of depression as a state? What if someone is having a day on which they are a bit *dramatic*? Those points are not bare nuances of a condition, they are major differences in what might be really happening to and with someone. If the data that a model is based on do not adequately represent reality, or measure robustly, specifically, and directly what one is intending to measure (in this exemplary case, depression), then the resulting model cannot be expected to deliver accurate, bias-free, meaningful inferences that advance our understanding of psychiatric disease or help us diagnose or treat it better. Compounded by the constraints outlined above (nosology, the consensual instead of an aetiological nature of psychiatric diagnoses, lack of biomarkers, difficulty of measuring the validity of predicted models), the validity of such models would be questionable as one would be trying to paint something of which one barely sees its shadow while not even painting it yourself but instructing someone else on how to paint it.

Although there are several publications showing algorithms predicting various psychiatric conditions from suicide to development of psychosis (Cohen et al., 2020; de Nijs et al., 2021; Koutsouleris et al., 2021), there is little to no evidence that they can be accurate, efficient, feasible and useful prospectively, let alone an implementation of those in psychiatric clinical practice to date. As most of those algorithms are based on small datasets, data from clinical trials or research settings, they are not likely to perform as well in clinical settings, as they are not representative of the real-life environment they are designed for. Moreover, even if they show acceptable accuracy, they are often inheritably biased and skewed in their predictions, with the potential of creating and reinforcing inequities in healthcare.

It can be contended that, given a substantial sample size and a diverse set of robust features—particularly encompassing data modalities like genetic or imaging data—AI has the potential to counterbalance the inherent limitations of self-report and observation-based data. By doing so, it could derive meaningful models, thereby advancing not only our comprehension but also the diagnosis, prediction, and treatment of psychiatric diseases. However, while this assertion is contingent upon future empirical validation, psychiatry should not overhype the potential of sophisticated computational methods to revolutionise itself if long-standing fundamental challenges of psychiatry are not overcome.

One solution to generating data that is more AI-friendly and robust would be using digital trackers such as wearables, social media and other digital mediums (Adler et al., 2022; Jacobson, Weingarden, & Wilhelm, 2019; Parziale & Mascalzoni, 2022). Such data can be very valuable due to its potentially high temporal resolution and continuity, thus being less prone to be flawed by temporary states in the psychological state of an individual. However, given that data on feelings, thoughts, perception, and behaviour of a human being are extremely sensitive, it is challenging to produce the data in large amounts by simultaneously ensuring that data privacy is not violated both from a technical and an ethical point of view.

4. Conclusion

Despite extensive research in this domain, computational psychiatry has led so far to scarce tangible results and suboptimal practical changes to the field of psychiatry (Huys et al., 2016). During this period of intense enthusiasm for AI, there is widespread speculation about its potential to transform the field of medicine, including psychiatry. Yet, if we take a pragmatic view, we must recognize that presently, there are hardly any practical applications or established practices in psychiatry that are significantly influenced or enhanced by computational psychiatry or AI. Despite high expectations for AI's capabilities, the reality is that we are still a considerable distance from witnessing a revolutionary impact on everyday clinical routines.

To quote Joel Paris' *Overdiagnosis in Psychiatry*, "while neuroscience has produced many exciting findings about normal brain functions, its application to psychiatry has been overhyped and remains only a promise. The beautiful (but artificially) coloured pictures provided by brain scans impress us. Yet one cannot predict, except in very general terms, thoughts, emotions or behaviour by what area of the brain lights up". (Paris, 2020).

Thomas Insel, reflecting on his time as the head of National Institute of Mental Health, stated: "I spent 13 years at NIMH really pushing on the neuroscience and genetics of mental disorders, and when I look back on that I realize that while I think I succeeded at getting lots of really cool papers published by cool scientists at fairly large costs, I don't think we moved the needle in reducing suicide, reducing hospitalizations, improving recovery for the tens of millions of people who have mental illness." (Regalado, 2015).

This chapter has argued that without tackling the fundamental problems posed by nosology, nature of data and reductionism, it might not be possible to evade a similar fate for computational psychiatry.

References

Adams, R. A., Huys, Q. J. M., & Roiser, J. P. (2016). Computational Psychiatry: Towards a mathematically informed understanding of mental illness. *Journal of Neurology, Neurosurgery, and Psychiatry, 87*, 53–63.

Adler, D. A., Wang, F., Mohr, D. C., Estrin, D., Livesey, C., & Choudhury, T. (2022). A call for open data to develop mental health digital biomarkers. *BJPsych Open, 8*, e58.

Anderson, C. (2008). *The end of theory: The data deluge makes the scientific method obsolete.* Wired,.

Angermeyer, M. C., Holzinger, A., Carta, M. G., & Schomerus, G. (2011). Biogenetic explanations and public acceptance of mental illness: Systematic review of population studies. *The British Journal of Psychiatry: The Journal of Mental Science, 199*, 367–372.

Baquet, C. R., Commiskey, P., Daniel Mullins, C., & Mishra, S. I. (2006). Recruitment and participation in clinical trials: Socio-demographic, rural/urban, and health care access predictors. *Cancer Detection and Prevention, 30*, 24–33.

Barron, D. ({C}n.d.).{C} Should mental disorders have names? ⟨https://blogs.scientificamerican.com/observations/should-mental-disorders-have-names/⟩ (accessed 14.10.23).

Beauchamp, T. L., & Childress, J. F. (2013). *Principles of biomedical ethics.* Oxford University Press.

Beck, A. T., Ward, C. H., Mendelson, M., Mock, J., & Erbauch, J. ({C}n.d.){C}. Beck depression inventory. *Archives of General Psychiatry* ⟨https://doi.org/10.1037/t00741–000⟩.

Bennett, M., Hayes, K., Kleczyk, E. J., & Mehta, R. (2022). Similarities and differences between machine learning and traditional advanced statistical modeling in healthcare analytics. *arXiv.*

Birnbaum, M. L., Ernala, S. K., Rizvi, A. F., Arenare, E., R Van Meter, A., De Choudhury, M., & Kane, J. M. (2019). Detecting relapse in youth with psychotic disorders utilizing patient-generated and patient-contributed digital data from Facebook. *NPJ Schizophrenia, 5*, 17.

Chan, D. (2010). *So why ask me? Are self-report data really that bad? Statistical and methodological myths and urban legends.* Routledge,329–356.

Chaudhury, D., Liu, H., & Han, M.-H. (2015). Neuronal correlates of depression. *Cellular and Molecular Life Sciences: CMLS, 72*, 4825–4848.

Christen, M., Domingo-Ferrer, J., Draganski, B., Spranger, T., & Walter, H. (2016). On the compatibility of big data driven research and informed consent: The example of the human brain project. In B. D. Mittelstadt, & L. Floridi (Eds.). *The ethics of biomedical big data* (pp. 199–218)Cham: Springer International Publishing.

Cohen, J., Wright-Berryman, J., Rohlfs, L., Wright, D., Campbell, M., Gingrich, D., ... Pestian, J. (2020). A feasibility study using a machine learning suicide risk prediction model based on open-ended interview language in adolescent therapy sessions. *International Journal of Environmental Research and Public Health, 17.* https://doi.org/10.3390/ijerph17218187.

Coveney, P. V., Dougherty, E. R., & Highfield, R. R. (2016). Big data need big theory too. *Philosophical Transactions: Mathematical, Physical and Engineering Sciences, 374.* https://doi.org/10.1098/rsta.2016.0153.

Cowgill, B., Dell'Acqua, F., Deng, S., Hsu, D., Verma, N., & Chaintreau, A. (2020). Biased programmers? Or biased data? A field experiment in operationalizing AI ethics. *arXiv.*

de Nijs, J., Burger, T. J., Janssen, R. J., Kia, S. M., van Opstal, D. P. J., de Koning, M. B., ... Schnack, H. G. (2021). Individualized prediction of three- and six-year outcomes of psychosis in a longitudinal multicenter study: A machine learning approach. *NPJ Schizophrenia, 7*, 34.

Dressel, J., & Farid, H. (2018). The accuracy, fairness, and limits of predicting recidivism. *Science Advances, 4*, eaao5580.

Floridi, L., Cowls, J., Beltrametti, M., Chatila, R., Chazerand, P., Dignum, V., ... Vayena, E. (2018). AI4People-an ethical framework for a good AI society: Opportunities, risks, principles, and recommendations. *Minds and Machine, 28*, 689–707.

Friston, K. (2023). Computational psychiatry: from synapses to sentience. *Molecular Psychiatry, 28*, 256–268.

Fuchs, T. (2010). Subjectivity and intersubjectivity in psychiatric diagnosis. *Psychopathology, 43*, 268–274.

Fuchs, T. (2006). Ethical issues in neuroscience. *Current Opinion in Psychiatry, 19*, 600–607.

Fusar-Poli, P., Tantardini, M., De Simone, S., Ramella-Cravaro, V., Oliver, D., Kingdon, J., ... McGuire, P. (2017). Deconstructing vulnerability for psychosis: Meta-analysis of environmental risk factors for psychosis in subjects at ultra high-risk. *European Psychiatry: The Journal of the Association of European Psychiatrists, 40*, 65–75.

Garber, M., & Arnold, R. M. (2006). Promoting the participation of minorities in research. *The American Journal of Bioethics: AJOB, 6*, W14–W20.

Gichoya, J. W., Banerjee, I., Bhimireddy, A. R., Burns, J. L., Celi, L. A., Chen, L.-C., ... Zhang, H. (2022). AI recognition of patient race in medical imaging: A modelling study. *Lancet Digital Health, 4*, e406–e414.

Guru Prasad, M. S., Gujjar, P., Naveen Kumar, H. N., Anand Kumar, M., & Chandrappa, S. (2023). Advances of cyber security in the healthcare domain for analyzing data. *Cyber trafficking, threat behavior, and malicious activity monitoring for healthcare organizations* (pp. 1–14) IGI Global.

Hassija, V., Chamola, V., Mahapatra, A., Singal, A., Goel, D., Huang, K., ... Hussain, A. (2023). Interpreting black-box models: A review on explainable artificial intelligence. *Cognitive Computing.* https://doi.org/10.1007/s12559-023-10179-8.

Health Ethics & Governance. (2021). Ethics and governance of artificial intelligence for health. ⟨https://www.who.int/publications/i/item/9789240029200⟩ (accessed 7.11.21).

Howe, E. G., Iii, & Elenberg, F. (2020). Ethical challenges posed by big data. *Innovations in Clinical Neuroscience, 17*, 24–30.

Huys, Q. J. M., Maia, T. V., & Frank, M. J. (2016). Computational psychiatry as a bridge from neuroscience to clinical applications. *Nature Neuroscience, 19*, 404–413.

Huys, Q. J. M., Moutoussis, M., & Williams, J. (2011). Are computational models of any use to psychiatry? *Neural Networks: The Official Journal of the International Neural Network Society, 24*, 544–551.

Huysmans, J., Baesens, B., & Vanthienen, J. (2006). Using rule extraction to improve the comprehensibility of predictive models. *Behavioral & Experimental Economics.* https://doi.org/10.2139/ssrn.961358.

Jacobs, M., Pradier, M. F., McCoy, T. H., Jr, Perlis, R. H., Doshi-Velez, F., & Gajos, K. Z. (2021). How machine-learning recommendations influence clinician treatment selections: The example of the antidepressant selection. *Translational Psychiatry, 11*, 108.

Jacobson, N. C., Weingarden, H., & Wilhelm, S. (2019). Digital biomarkers of mood disorders and symptom change. *npj Digital Medicine, 2*, 3.

Kant, I. (1787). Critique of pure reason.

Kendler, K. S. (2016). The nature of psychiatric disorders. *World Psychiatry: Official Journal of the World Psychiatric Association (WPA), 15*, 5–12.

Kirmayer, L. J., & Gold, I. (2011). *Re-socializing psychiatry. Critical neuroscience.* Oxford: Wiley-Blackwell,305–330.

Koutsouleris, N., Dwyer, D. B., Degenhardt, F., Maj, C., Urquijo-Castro, M. F., Sanfelici, R., ... PRONIA Consortium. (2021). Multimodal machine learning workflows for prediction of psychosis in patients with clinical high-risk syndromes and recent-onset depression. *JAMA Psychiatry, 78*, 195–209.

Lebowitz, M. S., & Ahn, W.-K. (2014). Effects of biological explanations for mental disorders on clinicians' empathy. *Proceedings of the National Academy of Sciences, 111*, 17786–17790.

Malhi, G. S., & Mann, J. J. (2018). Depression. *Lancet, 392*, 2299–2312.

Montague, P. R., Dolan, R. J., Friston, K. J., & Dayan, P. (2012). Computational psychiatry. *Trends in Cognitive Sciences, 16*, 72–80.

Nordgaard, J., Sass, L. A., & Parnas, J. (2013). The psychiatric interview: Validity, structure, and subjectivity. *European Archives of Psychiatry and Clinical Neuroscience, 263*, 353–364.

Paris, J. (2020). *Overdiagnosis in psychiatry: How modern psychiatry lost its way while creating a diagnosis for almost all of life's misfortunes.* Oxford University Press.

Parziale, A., & Mascalzoni, D. (2022). Digital biomarkers in psychiatric research: Data protection qualifications in a complex ecosystem. *Frontiers in Psychiatry, 13*, 873392.

Penfold, J. (2023). The growing risk of cyber attacks in the NHS. *British Journal of Healthcare Management, 29*, 5–7.

Regalado, A. (2015). *Why America's top mental health researcher joined alphabet.* MIT Technology Review.

Ríssola, E. A., Losada, D. E., & Crestani, F. (2021). A survey of computational methods for online mental state assessment on social media. *ACM Transactions on Computing for Healthcare, 2*, 1–31.

Robins, R. W., Chris Fraley, R., & Krueger, R. F. (2009). *Handbook of research methods in personality psychology.* Guilford Press.

Rutledge, R. B., Chekroud, A. M., & Huys, Q. J. (2019). Machine learning and big data in psychiatry: Toward clinical applications. *Current Opinion in Neurobiology, 55*, 152–159.

Sahin, D., Jessen, F., & Kambeitz, J. (2022). Algorithmic fairness in biomarker-based machine learning models to predict Alzheimer's dementia in individuals with mild cognitive impairment. *Alzheimer's & Dementia: The Journal of the Alzheimer's Association.*

Şahin, D., Kambeitz-Ilankovic, L., Wood, S., Dwyer, D., Upthegrove, R., ... Salokangas, R. PRONIA Study Group. (2023). Algorithmic fairness in precision psychiatry: analysis of prediction models in individuals at clinical high risk for psychosis. *British Journal of Psychiatry, 224*(2), 55–65.

Succi, S., & Coveney, P. V. (2019). Big data: The end of the scientific method? *Philosophical Transactions: Mathematical, Physical and Engineering Sciences, 377*, 20180145.

Sumathipala, A., Siribaddana, S., & Patel, V. (2004). Under-representation of developing countries in the research literature: Ethical issues arising from a survey of five leading medical journals. *BMC Medical Ethics, 5*, E5.

Suresh, H., & Guttag, J. (2021). *A framework for understanding sources of harm throughout the machine learning life cycle. Equity and access in algorithms, mechanisms, and optimization, EAAMO '21.* New York: Association for Computing Machinery,1–9.

Thorstad, R., & Wolff, P. (2019). Predicting future mental illness from social media: A big-data approach. *Behavior Research Methods, 51*, 1586–1600.

Uusitalo, S., Tuominen, J., & Arstila, V. (2021). Mapping out the philosophical questions of AI and clinical practice in diagnosing and treating mental disorders. *Journal of Evaluation in Clinical Practice, 27*, 478–484.

Van der Kolk, B. (2014). *The body keeps the score: Brain, mind, and body in the healing of trauma.* Viking,.

Wiese, W., & Friston, K. J. (2022). AI ethics in computational psychiatry: From the neuroscience of consciousness to the ethics of consciousness. *Behavioural Brain Research, 420*, 113704.

Zachar, P., & Kendler, K. S. (2017). The philosophy of nosology. *Annual Review of Clinical Psychology, 13*, 49–71.

> CHAPTER THREE

Computational psychiatry and digital phenotyping: Ethical and neurorights implications

José M. Muñoz[a,*], Diego Borbón[b], and Ana Maria Bezerra[c]

[a]Kavli Center for Ethics, Science, and the Public, University of California, Berkeley, CA, United States
[b]Center for Studies on Genetics and Law, Universidad Externado de Colombia, Bogotá, Colombia
[c]Unichristus School of Law, Unichristus, Fortaleza, CE, Brazil
*Corresponding author. e–mail address: jmmunoz@berkeley.edu

Contents

1. Introduction	48
2. Computational psychiatry and its ethical challenges	48
3. The case of digital phenotyping: From health applications to the consumer domain	51
4. Potential implications for neurorights	54
5. Final remarks: Towards the integration of different data levels	57
References	58

Abstract

Computational psychiatry is an interdisciplinary field that integrates and combines data at different levels and uses artificial intelligence models to understand the causes of psychiatric disorders, identify them individually, and improve diagnosis and treatment. One of the most important levels of analysis used in this discipline is that which integrates neural data, such as those obtained through functional magnetic resonance imaging and other brain imaging techniques. An especially promising application of computational psychiatry is digital phenotyping, which consists of the application of computational techniques to analyse cognitive, behavioural, motor and sleep data obtained with smartphones and other digital devices. Given the ease with which these data can be collected, digital phenotyping could be widely used in a relatively simple way, not only for psychiatric purposes but also for others related to user profiling or neuromarketing. In this chapter, we explore some ethical-legal implications—including those that fall within the framework of so-called neurorights—of computational psychiatry and digital phenotyping, especially in terms of privacy and algorithmic bias.

Developments in Neuroethics and Bioethics, Volume 7
ISSN 2589-2959, https://doi.org/10.1016/bs.dnb.2024.02.005
Copyright © 2024 Elsevier Inc. All rights are reserved, including those for text and data mining, AI training, and similar technologies.

1. Introduction

Computational psychiatry (CP) is an interdisciplinary field that integrates and combines data at different levels and uses artificial intelligence (AI) models to understand the causes of psychiatric disorders, identify them individually and improve their diagnosis and treatment. In doing so, CP inherits many of the typical dilemmas of AI ethics:

> Methods used in computational psychiatry (CP), such as deep learning, Bayesian modelling, or reinforcement learning, overlap with methods used in AI. Although the methods may be used for different aims, they can raise similar ethical issues. Hence, considerations from AI ethics are also relevant to CP. For instance, algorithms may produce unfair outcomes if their training data are biased and the possibility to collect and analyse personal data using algorithms raises issues of data ownership and protection. Furthermore, many applications of AI are not explicable, that is, it is often difficult or impossible to determine why an AI system yields a given outcome or who is accountable for the particular way in which an AI system works. Such immediate ethical concerns arise for applications of AI in general, but also for applications in mental healthcare and CP in particular. (Wiese & Friston, 2022: 1).

One of the most important levels of analysis used in CP is that which uses neural data, such as those obtained through functional magnetic resonance imaging and other brain imaging techniques (see Cohen et al., 2017). The integrated analysis of neural data in conjunction with behavioural and contextual data entails a novel yet challenging approach to the study of mental conditions and opens the path to an array of ethical and legal quandaries. In this chapter, we will explore some of these quandaries, especially in terms of privacy and algorithmic bias, regarding CP in general (Section 2) and digital phenotyping in particular (Section 3). We will also comment on possible pathways for legal debate about CP in the context of so-called neurorights (Section 4).

2. Computational psychiatry and its ethical challenges

CP aims, on a general level, for a better quality of life for patients. It does so by working with mathematical models and AI to automatically collect, analyse, compare and catalogue health data, with the intention of improving the understanding, diagnosis and treatment of mental disorders (Wiese & Friston, 2022).

Specifically, there are two approaches of CP that may complement each other for this purpose (Huys, Maia, & Frank, 2016): *data-driven*, which focuses on the extraction of data, such as symptoms description and medical exams results, and on the identification of relations and patterns between them; and *theory-driven*, which makes use of theoretical models and statistics in an attempt to better understand and establish prediction and prognosis for patients. Both approaches, whether they function top-down or bottom-up, involve the use of human data—mostly cognitive, behavioural and brain-related—at some point of their processes.

As this type of data has the potential to carry fundamental information about a person (Ienca et al., 2022), CP raises important ethical challenges, especially considering that it is an area that strongly benefits from advances in neurotechnology (see Goering et al., 2021). For example, there is the promised-to-be new generation of Apple's AirPods Sensor System (Patenly Apple, n.d.), intended to measure relevant biosignals by an individualised wearable device as simple as a headset. Accordingly, this kind of product may certainly help in controlling health aspects of an individual's life, but the reflections on the extent to which this control is necessary or even desired should not be ignored.

Indeed, it is feasible nowadays to precisely measure and monitor bio-signals such as temperature, weight, sleep and menstrual cycles, sexual activity, heartbeat rate and ovulation frequency in an active or passive way by simply using smartphones apps and tools. The possibility of easily registering more subjective and behavioural factors and inferring emotional states and personality inclinations from voice intonation, music preferences and the social media accounts that users follow is also worth mentioning.

These aspects, even more so if analysed together, are able to tell so much about a person, and as there are many ways for them to be collected and organised (Lin, 2022), it is increasingly easier and faster for professionals to both study and put into practice the materials inferred. However, these analyses may go beyond patients' medical-related conditions to reach their lives as a whole and, through crossing over and consequently *labelling* such data, even the lives of others. From that perspective, depending on whose hands those data are in, they can be used for purposes far from those initially and individually expected, raising concerns about privacy, pre-judice and consent-related decision-making.

In that sense, a practice as common as labelling holds significant rele-vance within human behaviour, proving indispensable for navigating a complex world. This is due to its capacity to provide us with the ability to comprehend reality without needing exhaustive analysis whenever we encounter new situations. However, the challenge lies in the foundation upon which our categorisations are built, as we frequently rely on readily available and familiar information to slot individuals into predefined boxes (Tversky & Kahneman, 1973). There are logical reasons for doing so, but depending on the context, these reasons might not consistently align with objective truth; rather, they can sometimes resemble more of a leap of faith. Moreover, the possibility of inference errors increases when this leap of faith is built on high-scale data collection, such as those made possible by CP and automated by AI tools such as algorithms, possibly leading to biased profiling.

Merely possessing an extensive and/or more accurate database would initially yield positive outcomes. It would signal a higher likelihood of an inference aligning with reality. As the volume of collected data expands, the tendency is for the resulting categorisations to become more robust and reliable. This principle forms the foundation of CP methodologies, making it possible to level up diagnosis prediction and treatment plans, but it also introduces the potential for a stigmatised reductionism by automatising the process of labelling patients typical or atypical, and oversimplifying decisions on how to deal with their brain-related health issues (Uusitalo, Tuominen, & Arstila, 2021). Furthermore, this categorisation may be used for neuromarketing, which makes use of personalised nudging strategies towards a sale, most often without the user's explicitly conscious consent.

As mentioned, although the use of automatically categorised data within the scope of CP is mainly understood as a promising advance because it aims for a better quality of life for patients—an objective it often successfully achieves[1]—the associated ethical challenges need to be evaluated, as they have a clear effect when appropriately addressed. The more these reflections transcend the boundaries of the academic world, the greater the likelihood that international policymakers will pay attention to these matters. Such attention results in the establishment of clear guidelines and limitations for data users,[2] which leads to companies complying and worrying about those ethical challenges, in the way TikTok (2023) recently did in the matters of informed consent.

To properly face the known and yet-to-be-known ethical concerns, it is also necessary to take a specific and deep look at the different applications of neurotechnology. With these considerations in mind, we acknowledge

---

[1] For instance, see the research direction on how CP—especially speech measurement—can contribute to unbiased diagnosis and treatment in multicultural societies like Brazil (Mota et al., 2022).

[2] See, for example, UNESCO (2023) and Garden and Winickoff (2018).

that an interdisciplinary approach to CP is indispensable, considering that it allows a more complete understanding of how to deal with its challenges, points to a more reliable management of data and finds the balance between its benefits and risks.

3. The case of digital phenotyping: From health applications to the consumer domain

An especially promising application of CP is *digital phenotyping*. This technique, created by the Onnela Lab (n.d.) at the Harvard Chan School of Public Health, is defined as 'moment-by-moment quantification of the individual-level human phenotype in-situ using data from smartphones and other personal digital devices' (Torous, Kiang, Lorme, & Onnela, 2016: 2). Through a platform called Beiwe and by using computational techniques, this team has developed a methodology to analyse cognitive, behavioural, motor and sleep data to obtain a complete profile of patients at various levels.

Digital phenotyping can also be integrated with brain imaging neuro-technologies (see Camacho, Brady, Lizano, Keshavan, & Torous, 2021) and other tools employed in the medical field:

> *The data from these [personal digital] devices can be combined with electronic medical records and with molecular and neuroimaging data. In this sense, digital phenotyping can be viewed as a variant of deep phenotyping. Digital phenotyping is also closely aligned with the goals of precision medicine, which links new types of phenotypic data with genome data in order to identify potential connections between disease subtypes and their genetic variations.*
> (Torous et al., 2016: 2).

Due to this successful integration of data at multiple levels, digital phenotyping has shown great potential for the diagnosis and monitoring of several brain and mental disorders (Onnela, 2021; Straczkiewicz, James, & Onnela, 2021; Torous, Gershon, Hays, Onnela, & Baker, 2019), including schizophrenia (Torous, Firth, Mueller, Onnela, & Baker, 2017), amyotrophic lateral sclerosis (Berry et al., 2019), spine disease (Cote, Barnett, Onnela, & Smith, 2019) and depression and anxiety (Aledavood et al., 2019). Ultimately, '[f]or neurology, which has required expensive, clinic-based assessments of cognitive performance, digital phenotyping offers an inexpensive, ecological assessment of function under real-world conditions' (Martinez-Martin, Insel, Dagum, Greely, & Cho, 2018: 1).

However, digital phenotyping techniques are not limited to clinical use, since they can also be applied in areas as diverse as criminal law, employee recruitment by companies, education, research and even the *consumer domain* (Martinez-Martin et al., 2018). Indeed, given its efficacy in the collection and analysis of users' data, digital phenotyping can be used for profiling or marketing purposes in a relatively simple way. All these applications raise important ethical-legal questions, which Martinez-Martin et al. (2018) group into four main axes:

- *Accountability*, since no current regulatory framework seems to encompass this kind of methodology;
- *Personal data protection*, to the extent that users tend to be permissive with the use of their digital data because they do not consider them health data or are not aware of their sensitivity or identifiability;
- *Transparency* regarding the what, when, and how in the gathering of personal data;
- *Informed consent*, whose framework would need to be updated to ensure that patients and users are aware of the implications of these new technologies.

Other scholars have emphasised the fact that there are a number of additional ethical challenges that are not usually considered in the literature on the topic:

Firstly, given the impact of such research on the lives of individuals and patients, it seems essential to conduct empirical studies through field research on the dynamics of digital phenotyping technology seen as a social artifact; on the psychosociological influences of these technologies on the individual, inter-personal, societal, and global level; and on the definition of the boundaries between well-being and health and illness. In the same vein, the physical and psychological repercussions of wearable devices need to be examined from the point of view of various profiles of users, based on their experience within their specific social and cultural contexts. Tomičić, Malešević, and Čartolovni (2022: 22).

This last reference to wearable devices connects with another critical ethical challenge: the potential integration of digital phenotyping with neurotechnologies, particularly with those known as direct-to-consumer (DTC) neurotechnological devices (see Ienca & Vayena, 2019; Wexler & Reiner, 2019). The production of DTC neurotechnologies is quickly escalating today and they can be easily purchased in physical and online stores. Widespread use of these devices in the future could bring with it unprecedented ethical and regulatory challenges, including those associated with digital phenotyping and user profiling.

This integration between digital phenotyping and DTC neuro-technologies is likely to be promising for neuromarketing departments around the world to improve companies' abilities to offer personalised content to their users. An illustrative example of how neurotechnologies and digital devices for personal use can provide valuable integrated information for companies is a recent study conducted by Spotify (2022), in which steady state topography was employed to measure cerebral electrical activity generated by several users while they listened to different songs. This research made it possible to infer the emotional intensity of these users as a reaction to musical content, leading the company to better know which songs tend to be the most liked according to specific profiles.

Commercially speaking, offering highly personalised content to users seems of great advantage not only for companies but also for users, who can benefit from having useful or enjoyable content at their disposal. However, this degree of personalisation also poses some important risks, since '[i]n some commercial contexts, people may have lowered expectations of privacy or be willing to share some personal data in order to receive a perceived benefit' (Martinez-Martin et al., 2018: 3). What is more, various neurotechnological companies are enforcing privacy policies that allow them to share brain data with outside parties. By doing so, these companies are exposing users to a high risk of these data being used by third parties to perform profiling and phenotyping for unknown, unauthorised purposes that may, in some cases, even erode freedom of conscience and personal dignity.

Much can be learned about a person from their brain data. Recent research seems to show, for example, that it is possible to predict, with a relevant percentage of success, a person's political ideology from the mere study of their brain data (Yang, Wilson, Lu, & Cranmer, 2022). The transfer or sale to third parties of these types of data in combination with personal data collected from digital devices is a sensitive matter that regards issues such as dignity and honour. For instance, think of the possibility of a third party performing a personal profile for a user that includes their emotional reactions (derived from their brain data) while watching some kind of delicate or extreme audiovisual content (e.g., a violent crime, a pornographic scene) on a digital platform. Additional challenges may arise in the case of vulnerable groups such as minors, the elderly, women, ethnic and racial minorities and sexually diverse people.

While the combination of digital phenotyping and neurotechnology could be extremely beneficial for patients, users and consumers, all necessary ethical analysis and corresponding regulatory measures should be taken in order to avoid these methodologies being used (in the wrong hands) as means to aggravate existing problems of so-called surveillance capitalism (see Zuboff, 2019). As stressed by Martinez-Martin and Char (2018: 68) in the case of healthcare, '[t]aking the time now to address the direction of digital medicine will help ensure that the physician–patient relationship does not come to resemble Big Brother'.

4. Potential implications for neurorights

The ascendancy of CP and neurotechnologies opens a Pandora's box of ethical, legal and social quandaries, warranting serious considerations through the lens of so-called neurorights. In this way, the deep effects of digital phenotyping on the quantification of behavioural, biological and sensitive data at the individual level, together with the integration of neurotechnologies, raise important questions about *mental privacy*. As has been well-pointed out by Ienca et al. (2022: 3), 'brain data can be combined with non-neural contextual data, such as voice recordings, smartphone usage data [obtained, for example, through digital phenotyping] or neuropsychological assessments, that can be used to support inferences about mental processes in a broader sense'.

Recently, Ienca and Malgieri (2022) have introduced the concept of *mental data*, that is, data that allow the inference of an individual's mental states. According to these authors, mental data are not entirely equivalent to brain/neural data nor behavioural data, since part of both kinds of data do not provide information that allows the inference of mental states. However, some neural data and some behavioural data are mental data as long as they are used to make such inference. Thus, by capturing and analysing both neural and behavioural data, the confluence of CP—including digital phenotyping—and neurotechnologies might enable an association not only with neurophysiological states but also with subjective, emotional and volitional states.

A second crucial implication revolves around *algorithmic bias*. The composite data from digital phenotyping and neurotechnologies are subject to computational analysis, which inevitably involves AI. These algorithms may inherit or exacerbate societal biases, thereby compromising the equitable treatment of individuals, especially those belonging to vulnerable groups and underrepresented minorities. Data harvested from neurotechnologies and behavioural observations could inadvertently be biased or misrepresentative, and when analysed collectively, they could magnify pre-existing stereotypes and social inequalities. Since bad predictions and unfair decisions could have serious effects in the real world (Schulz & Dayan, 2020), the introduction of legislative frameworks must include specific clauses aimed at detecting, preventing and rectifying algorithmic biases.

According to Wiese (2021), there are important risks such as: (1) increasing social stigma; (2) biases inherent in the training data, which could result in ineffective or even harmful diagnoses and treatments; (3) violations of data privacy when large datasets are used and when information is shared across different institutions. For these reasons, AI ethics is extremely relevant for CP to prevent unfair outcomes and protect personal data (Wiese & Friston, 2022).

The multifaceted dimensions of ethical, legal and social issues unearthed by CP, digital phenotyping and neurotechnologies lay the groundwork for an important debate on *neurorights*. This discourse can serve as a robust framework for the conceptualisation and enactment of legal safeguards specifically targeting these emergent technologies. Ienca (2021:1) has recently defined neurorights 'as the ethical, legal, social, or natural principles of freedom or entitlement related to a person's cerebral and mental domain; that is, the fundamental normative rules for the protection and preservation of the human brain and mind'. This concept has become important in legal domains and has to undergo academic scrutiny as it transcends traditional human rights doctrines and anticipates the specialised nature of challenges such as mental data privacy and algorithmic bias.

In light of the foregoing discussions, the construct of neurorights might provide a new legal and ethical scaffolding to navigate these challenges. Predicated on seminal works by Ienca and Andorno (2017) as well as by Yuste et al. (2017), neurorights encompass a set of proposed fundamental human rights to address the novel quandaries engendered by neuroscientific and technological developments. These rights function as instruments of the human rights landscape, where technological capabilities may exceed the protective coverage offered by traditional conventional and constitutional norms.

Among the specific neurorights proposed, mental privacy (Ienca & Andorno, 2017; Yuste et al., 2017) and protection against bias (Yuste et al., 2017) serve as cornerstones to mitigate the ethical and legal challenges arising from neurotechnologies. On one hand, the right to mental privacy not only aims to safeguard brain data but also seeks to establish barriers against unauthorised data collection, storage and utilisation of neural information and the mental data that could be inferred from raw brain data (Ienca & Andorno, 2017; Ienca, 2021). On the other hand, the right to protection from algorithmic bias aims to mitigate the spectrum of harms that are engendered by biased AI systems (Ienca, 2021; Yuste, Genser, & Herrmann, 2021).[3]

The landscape of neurorights is increasingly being formalised through regulations in countries like Argentina, Brazil, Chile, France, Mexico and Spain, while international institutions including the Latin American Parliament (Parlatino), the Organization of American States (OAS) and the United Nations have also taken pivotal steps towards recognising and legislating neurorights (see Muñoz & Borbón, 2023). The Latin American Parliament (Parlatino, 2023) has launched a model law on neurorights that incorporates both protection against bias and mental privacy. Additionally, principles 1, 3 and 4 of the Inter-American Declaration of Principles Regarding Neuroscience, Neurotechnologies and Human Rights by the OAS Inter-American Juridical Committee (2023) tend to protect the privacy of neural activity as sensitive personal data and require express and informed consent regarding its access and use.

As a real case example, on 9 August 2023, the Third Chamber of the Supreme Court of Chile, Third Chamber (2023) ruled in favour of a constitutional protection action brought by ex-Senator Guido Girardi, a supporter of neurorights in his country. The court said that a wireless electroencephalography device had violated the express consent to use neural data. In this sense, for having violated the constitutional guarantees contained in numerals 1 and 4 of Article 19 of the Chilean Constitution, which refer to physical and mental integrity and the right to privacy, the Supreme Court ordered the complete elimination of said data.

Considering all these advancements in the neurorights framework,[4] the advent of CP and digital phenotyping presents both opportunities and

---

[3] Some other neurorights proposed by Ienca and Andorno (2017) are the rights to cognitive liberty, mental integrity and psychological continuity. Yuste et al. (2021) and the NeuroRights Foundation (n.d.), for their part, propose the rights to free will, personal identity and fair access to mental augmentation.

[4] Be that as it may, it cannot be ignored that neurorights proposals have received important criticisms. Bublitz (2022) alerts to the risk that neurorights might be rather vague or redundant and suggests that some neurorights initiatives lack serious legal bases and contribute to a rights inflation that could undermine the robustness of current human rights paradigms. Fins (2022), on the other hand, is concerned that the Chilean reform on neurorights could inhibit medical research and pose ethical dilemmas, particularly in the area of disorders of consciousness, and he concludes that the aforementioned reform was vague and premature and that, consequently, it should not be imitated by any other jurisdiction. Considering these criticisms, it is important to stress that more academic and political discussion is still necessary to advance regulations that address the challenges presented by neurotechnological advances and their clinical and commercial applications.

challenges. On one hand, these emerging technologies offer the promise of enhanced diagnostic capabilities and personalised mental health interventions. However, due to their mixability with brain data, they also exacerbate existing concerns surrounding mental privacy and algorithmic bias. Given these complexities, there might be a pressing need for an advanced ethical and legal infrastructure that is capable of addressing the challenges posed by these technological advancements. The implications of CP and digital phenotyping require that policymakers, ethicists and clinicians have truly interdisciplinary discussions.

In terms of brain data, a group of scholars have recently suggested a governance framework consisting of 'four primary areas of regulatory intervention: binding regulation, ethics and soft law, responsible innovation, and human rights' (Ienca et al., 2022: 8). First, binding regulations should focus on stringent data protection measures, user consent and limitations on nonmedical and military applications. Second, ethical guidelines 'should extend beyond mere rule-compliance and promote the respectful use of brain data' (Ienca et al., 2022: 11). Third, responsible research and innovation should be based on emphasising the importance of safety, scientific validity, accountability and transparency, along with the implementation of advanced privacy-preserving technologies and continuous risk assessment. Fourth, Ienca et al. (2022: 11–12) conclude that '[h]uman rights inform legislation, ethical guidelines and societal norms across the globe, and thus offer an international normative framework where brain data protection needs to be embedded'.

5. Final remarks: Towards the integration of different data levels

The convergence of AI and neurotechnology tools (and data) will be an inevitable trend in the years to come, not only in the context of mental health but also in a variety of other areas such as neuromarketing, digital platforms, DTC neurotechnologies and criminal law. While we have focused in this chapter on the specific convergence between behavioural data and neural data, the integration of data from different organic levels includes other domains and disciplines, as demonstrated by the advent of *multi-omics*, a new field of inquiry that 'aims to combine two or more omics datasets to aid in data analysis, visualisation and interpretation to determine the mechanism of a biological process' (Krassowski, Das, Sahu, & Misra, 2020: 1). This field, almost by definition, involves the extraction and analysis of vast amounts of data:

> In the last decade, the application of different individual omic studies (e.g., genomics, epigenomics, transcriptomics, proteomics, metagenomics) that aimed at understanding a particular problem in human disease, agriculture, plant science, microbiology, and the environment have been successful to a great extent. These studies generate a plethora of data, which, with careful integration under a suitable statistical and mathematical framework, can help to solve broader queries pertaining to basic and applied areas of biology. (Krassowski et al., 2020: 1).

Nevertheless, this multi-omics data outbreak brings with it a variety of ethical challenges related to privacy in terms of reidentification of subjects and the sensitivity of the information obtained (Dupras & Bunnik, 2021).

Ultimately, the challenges for privacy and biases posed by the integration of different levels of data, both in CP and multi-omics, should be faced from different deontological perspectives, not only those based on respecting rights (such as neurorights) but also those working on the design of obligations—in the form, for example, of *neuroduties* (Echarte, 2021: 142) or a *technocratic oath* (Álamos et al., 2022)—as well as market self-regulation standards.

References

Álamos, M. F., Kausel, L., Baselga-Garriga, C., Ramos, P., Aboitiz, F., Uribe-Etxebarria, X., & Yuste, R. (2022). A technocratic oath. In P. López-Silva, & L. Valera (Eds.). *Protecting the mind* (pp. 163–174)Cham, Switzerland: Springer. https://doi.org/10.1007/978-3-030-94032-4_14.

Aledavood, T., Torous, J., Triana, A. M., Naslund, J. A., Onnela, J.-P., & Keshavan, M. (2019). Smartphone-based tracking of sleep in depression, anxiety, and psychotic disorders. *Current Psychiatry Reports, 21*, 49. https://doi.org/10.1007/s11920-019-1043-y.

Berry, J. D., Paganoni, S., Carlson, K., Burke, K., Weber, H., Staples, P., ... Onnela, J.-P. (2019). Design and results of a smartphone-based digital phenotyping study to quantify ALS progression. *Annals of Clinical and Translational Neurology, 6*, 873–881. https://doi.org/10.1002/acn3.770.

Bublitz, J. C. (2022). Novel neurorights: From nonsense to substance. *Neuroethics, 15*, 7. https://doi.org/10.1007/s12152-022-09481-3.

Camacho, E., Brady, R. O., Lizano, P., Keshavan, M., & Torous, J. (2021). Advancing translational research through the interface of digital phenotyping and neuroimaging: A narrative review. *Biomarkers in Neuropsychiatry, 4*, 100032. https://doi.org/10.1016/j.bionps.2021.100032.

Cohen, J. D., Daw, N., Engelhardt, B., Hasson, U., Li, K., Niv, Y., ... Willke, T. L. (2017). Computational approaches to fMRI analysis. *Nature Neuroscience, 20*, 304–313. https://doi.org/10.1038/nn.4499.

Cote, D. J., Barnett, I., Onnela, J.-P., & Smith, T. R. (2019). Digital phenotyping in patients with spine disease: A novel approach to quantifying mobility and quality of life. *World Neurosurgery, 126*, e241–e249. https://doi.org/10.1016/j.wneu.2019.01.297.

Dupras, C., & Bunnik, E. M. (2021). Toward a framework for assessing privacy risks in multi-omic research and databases. *The American Journal of Bioethics, 21*(12), 46–64. https://doi.org/10.1080/15265161.2020.1863516.

Echarte, L. E. (2021). De la neuroética al neo-romanticismo: La respuesta de Aldous Huxley a las propuestas actuales para la regulación ética y legal de la neurociencia. *Scio, 21*, 113–148. https://doi.org/10.46583/scio_2021.21.783.

Fins, J. J. (2022). The unintended consequences of Chile's neurorights constitutional reform: Moving beyond negative rights to capabilities. *Neuroethics, 15*, 26. https://doi.org/10.1007/s12152-022-09504-z.

Garden, H., & Winickoff, D. (2018). Issues in neurotechnology governance. *OECD Science, Technology and Industry Working Papers, 2018, 11*. https://doi.org/10.1787/c3256cc6-en.

Goering, S., Klein, E., Specker Sullivan, L., Wexler, A., Agüera y Arcas, B., Bi, G., ... Yuste, R. (2021). Recommendations for responsible development and application of neuro-technologies. *Neuroethics, 14*(3), 365–386. https://doi.org/10.1007/s12152-021-09468-6.

Huys, Q. J., Maia, T. V., & Frank, M. J. (2016). Computational psychiatry as a bridge from neuroscience to clinical applications. *Nature Neuroscience, 19*(3), 404–413. https://doi.org/10.1038/nn.4238.

Ienca, M. (2021). On neurorights. *Frontiers in Human Neuroscience, 15*, 701258. https://doi.org/10.3389/fnhum.2021.701258.

Ienca, M., & Andorno, R. (2017). Towards new human rights in the age of neuroscience and neurotechnology. *Life Sciences, Society and Policy, 13*, 5. https://doi.org/10.1186/s40504-017-0050-1.

Ienca, M., Fins, J. J., Jox, R. J., Jotterand, F., Voeneky, S., Andorno, R., ... Kellmeyer, P. (2022). Towards a governance framework for brain data. *Neuroethics, 15*, 20. https://doi.org/10.1007/s12152-022-09498-8.

Ienca, M., & Malgieri, G. (2022). Mental data protection and the GDPR. *Journal of Law and the Biosciences, 9*(1), lsac006. https://doi.org/10.1093/jlb/lsac006.

Ienca, M., & Vayena, E. (2019). Direct-to-consumer neurotechnology: What is it and what is it for? *AJOB Neuroscience, 10*(4), 149–151. https://doi.org/10.1080/21507740.2019.1668493.

Inter-American Juridical Committee. (2023). *Inter-American declaration of principles regarding neuroscience, neurotechnologies, and human rights (CJI/RES. 281 [CII-O/23] corr.1). Organization of American States.* https://www.oas.org/en/sla/iajc/docs/CJI-RES_281_CII-O-23_corr1_ENG.pdf.

Krassowski, M., Das, V., Sahu, S. K., & Misra, B. B. (2020). State of the field in multi-omics research: From computational needs to data mining and sharing. *Frontiers in Genetics, 11*, 610798. https://doi.org/10.3389/fgene.2020.610798.

Lin, B. (2022). Computational inference in cognitive science: Operational, societal and ethical considerations. *arXiv, 2210*, 13526. https://doi.org/10.48550/arXiv.2210.13526.

Martinez-Martin, N., & Char, D. (2018). Surveillance and digital health. *The American Journal of Bioethics, 18*(9), 67–68. https://doi.org/10.1080/15265161.2018.1498954.

Martinez-Martin, N., Insel, T. R., Dagum, P., Greely, H. T., & Cho, M. K. (2018). Data mining for health: Staking out the ethical territory of digital phenotyping. *npj Digital Medicine, 1*, 68. https://doi.org/10.1038/s41746-018-0075-8.

Mota, N. B., Pimenta, J., Tavares, M., Palmeira, L., Loch, A. A., Hedin-Pereira, C., & Dias, E. C. (2022). A Brazilian bottom-up strategy to address mental health in a diverse population over a large territorial area—An inspiration for the use of digital mental health. *Psychiatry Research, 311*, 114477. https://doi.org/10.1016/j.psychres.2022.114477.

Muñoz, J. M., & Borbón, D. (2023). Equal access to mental augmentation: Should it be a fundamental right? *Brain Stimulation, 16*(4), 1094–1096. https://doi.org/10.1016/j.brs.2023.05.003.

NeuroRights Foundation. (n.d.). *Mission: The five NeuroRights.* https://neurorightsfoundation. org/mission.

Onnela, J.-P. (2021). Opportunities and challenges in the collection and analysis of digital phenotyping data. *Neuropsychopharmacology: Official Publication of the American College of Neuropsychopharmacology, 46*, 45–54. https://doi.org/10.1038/s41386-020-0771-3.

Onnela Lab. (n.d.). *Research areas.* Harvard Chan School of Public Health. https://www. hsph.harvard.edu/onnela-lab/research/.

Parlatino. (2023). *Ley Modelo de Neuroderechos para América Latina y El Caribe.* https:// parlatino.org/wp-content/uploads/2017/09/leym-neuroderechos-7-3-2023.pdf.

Patently Apple. (n.d.). *Apple invents a next-generation AirPods Sensor System that could measure biosignals and electrical activity of a user's brain.* https://www.patentlyapple.com/2023/07/ apple-invents-a-next-generation-airpods-sensor-system-that-could-measure-biosignals-and-electrical-activity-of-a-users-brain.html.

Schulz, E., & Dayan, P. (2020). Computational psychiatry for computers. *iScience, 23*(12), 101772. https://doi.org/10.1016/j.isci.2020.101772.

Spotify. (2022). *Sonic science: Your brain on music.* https://ads.spotify.com/en-US/news-and-insights/sonic-science-your-brain-on-music/.

Straczkiewicz, M., James, P., & Onnela, J.-P. (2021). A systematic review of smartphone-based human activity recognition methods for health research. *npj Digital Medicine, 4*, 148. https://doi.org/10.1038/s41746-021-00514-4.

Supreme Court of Chile, Third Chamber. (2023). *Sentence No. 105065–2023, Page No. 217225-2023.*

TikTok. (2023). *An update on fulfilling our commitments under the Digital Services Act.* https:// newsroom.tiktok.com/en-eu/fulfilling-commitments-dsa-update.

Tomičić, A., Malešević, A., & Čartolovni, A. (2022). Ethical, legal and social issues of digital phenotyping as a future solution for present-day challenges: A scoping review. *Science and Engineering Ethics, 28*(1), https://doi.org/10.1007/s11948-021-00354-1.

Torous, J., Firth, J., Mueller, N., Onnela, J.-P., & Baker, J. T. (2017). Methodology and reporting of mobile health and smartphone application studies for schizophrenia. *Harvard Review of Psychiatry, 25*(3), 146–154. https://doi.org/10.1097/HRP.0000000000000133.

Torous, J., Gershon, A., Hays, R., Onnela, J.-P., & Baker, J. T. (2019). Digital phenotyping for the busy psychiatrist: Clinical implications and relevance. *Psychiatric Annals, 49*(5), 196–201. https://doi.org/10.3928/00485713-20190417-01.

Torous, J., Kiang, M., Lorme, J., & Onnela, J. (2016). New tools for new research in psychiatry: A scalable and customizable platform to empower data driven smartphone research. *JMIR Mental Health, 3*(2), e16. https://doi.org/10.2196/mental.5165.

Tversky, A., & Kahneman, D. (1973). Availability: A heuristic for judging frequency and probability. *Cognitive Psychology, 5*(2), 207–232. https://doi.org/10.1016/0010-0285(73)90033-9.

UNESCO. (2023). *The risks and challenges of neurotechnologies for human rights.* United Nations. https://unesdoc.unesco.org/ark:/48223/pf0000384185.

Uusitalo, S., Tuominen, J., & Arstila, V. (2021). Mapping out the philosophical questions of AI and clinical practice in diagnosing and treating mental disorders. *Journal of Evaluation in Clinical Practice, 27*(3), 478–484. https://doi.org/10.1111/jep.13485.

Wexler, A., & Reiner, P. B. (2019). Oversight of direct-to-consumer neurotechnologies. *Science (New York, N. Y.), 363*(6424), 234–235. https://doi.org/10.1126/science.aav0223.

Wiese, W. (2021). Von der KI-Ethik zur Bewusstseinsethik: Ethische Aspekte der Computational Psychiatry. *Psychiatrische Praxis, 48*(S 01), S21–S25. https://doi.org/10. 1055/a-1369-2824.

Wiese, W., & Friston, K. J. (2022). AI ethics in computational psychiatry: From the neuroscience of consciousness to the ethics of consciousness. *Behavioural Brain Research, 420*, 113704. https://doi.org/10.1016/j.bbr.2021.113704.

Yang, S. E., Wilson, J. D., Lu, Z.-L., & Cranmer, S. (2022). Functional connectivity signatures of political ideology. *PNAS Nexus, 1*(3), pgac066. https://doi.org/10.1093/pnasnexus/pgac066.

Yuste, R., Genser, J., & Herrmann, S. (2021). It's time for neuro-rights. *Horizons, 18,* 154–165. https://www.cirsd.org/en/horizons/horizons-winter-2021-issue-no-18/its-time-for-neuro–rights.

Yuste, R., Goering, S., Agüera y Arcas, B., Bi, G., Carmena, J. M., Carter, A., ... Wolpaw, J. (2017). Four ethical priorities for neurotechnologies and AI. *Nature, 551,* 159–163. https://doi.org/10.1038/551159a.

Zuboff, S. (2019). *The age of surveillance capitalism.* New York: PublicAffairs.

SECTION 2

Neuroscience for AI

CHAPTER FOUR

Neuroscience for AI: The importance of theory of mind

Christelle Langley[a,*], Fabio Cuzzolin[b], and Barbara J. Sahakian[a]
[a]Department of Psychiatry, University of Cambridge, Cambridge, United Kingdom
[b]School of Engineering, Computing and Mathematics, Oxford Brookes University, Oxford, United Kingdom
*Corresponding author. e-mail address: cl798@medschl.cam.ac.uk

Contents

1. Introduction	66
2. *Hot* and *cold* cognition	67
3. *Hot* and *cold* cognition in theory of mind	71
4. *Hot* and *cold* cognition and theory of mind in AI	73
5. Limitations and ethical concerns to theory of mind in AI	76
6. Conclusions	78
Acknowledgements	79
References	79

Abstract

Understanding Theory of Mind is challenging as this psychological domain can be viewed as a complex holistic process that can be decomposed into a number of *hot* and *cold* cognitive processes. *Cold* cognitive processes are non-emotional, whereas *hot* cognitive processes are both social and emotional. *Cold* cognition includes working memory, cognitive flexibility and if-then inferential logic and planning, processes which are used in non-social contexts, but which are often components in neuropsychological tests of Theory of Mind. In social situations, individuals use social cognition to process, remember and use information to explain and predict other people's behaviour, as well as their own. Therefore, strategic behaviour for goal achievement involving others often relies on an interaction between *hot* and *cold* cognition. Similarly, for goal achievement in artificial intelligence (AI) such as the robust performance of autonomous cars, or in the case of therapeutic interactions with humans, it is important to have both *cold* cognitive processes and *hot* cognitive processes. This chapter will predominantly address *hot* cognitive processes, which still require development in AI, compared to cold processes that are already well-established, and the underlying neural networks. In addition, we will cover how the understanding of hot cognitive processes in humans might inform the integration into AI models to more successfully mimic the human brain and enhance AI-human interactions. Finally, we also discuss the importance of an integrated and interdisciplinary approach to AI models as well as the growing ethical issues in brain-inspired AI.

1. Introduction

Theory of mind (ToM) is a key component of human cognition and refers to the ability to attribute mental states, such as beliefs, desires, and intentions, to oneself and others, and to use that information to understand and predict behaviour. In humans, it is a key component of social interaction and communication. ToM is more broadly incorporated as a key component of social cognition and involves both *hot* and *cold* cognitive mechanisms. The term ToM was first used by Premack and Woodruff (1978) in their attempt to determine whether a chimpanzee had such abilities. Subsequent research suggested a social brain hypothesis, where authors argued that, from an evolutionary standpoint, having good ToM abilities would be beneficial for increasingly complex social environments (Brothers, 1990). The development of ToM is typically observed in children around the age of four or five, and it is thought to be related to the growth of certain brain regions involved in social cognition. Some of the key subcomponents, such as making eye contact with parents and others, start much earlier.

Human societies are increasing reliant on technologies such as artificial intelligence (AI). Indeed, there have been great advancements in recent years with machine learning (ML), deep learning (DL) and reinforcement learning (RL) models. For example, the computer program AlphaGo consistently and famously beat the human world champion at Go, providing new insights on successful game strategies (Cuzzolin, Morelli, Cirstea, & Sahakian, 2020). Similarly, ChatGPT, a recently developed language model that is able to respond to user written prompts, has stunned with its ability to manipulate natural language. At the time of writing, ChatGPT currently has over 100 million users, and its website currently records 1 billion visitors per month (Duarte, 2023). Despite their impressive feats, even state-of-the-art AI models still miss a key component of human cognition: theory of mind.

In recent years, there has been growing interest in developing AI systems that can exhibit ToM (Conati, 2002; Gonzalez & Chang, 2021). The goal is to create machines that can not only understand human behaviour, but also interact with humans in a more natural and intuitive way. This is particularly important for human–robot interaction applications such as social robotics, virtual assistants, and autonomous vehicles. One approach to building ToM into AI systems is to use machine learning algorithms to analyse large datasets of human behaviour and infer underlying mental states. Cognitive and neuroimaging studies have examined ToM abilities, in particular research on individuals with autism spectrum disorder (ASD), where a ToM deficit is present, has particularly contributed to our understanding of human ToM. Given the knowledge that has been acquired about human ToM through cognitive science and neuroscience it could prove advantageous for machine ToM to utilise this knowledge. Another approach is to use reinforcement learning (Jara-Ettinger, 2019), where the AI system learns through trial and error by interacting with humans in a simulated environment. In this way, the machine can learn to recognize patterns in human behaviour and adjust its own behaviour accordingly.

While there has been some progress in developing AI systems with ToM, many challenges remain (Langley, Cirstea, Cuzzolin, & Sahakian, 2022). One major limitation is the need for large amounts of data to train these models, as well as the difficulty of creating realistic and complex simulations of human behaviour. Additionally, there are ethical concerns around the use of these systems. If these systems have ToM, will they be able to manipulate us by detecting our motivations, needs and desires? Overall, however, the development of AI with ToM has the potential to revolutionize the way we interact with machines and each other. By building machines that can understand and predict human behaviour, we can create more natural and intuitive interfaces that will make our lives easier and more enjoyable.

In this chapter, we aim to provide a better understanding of *hot* and *cold* cognitive mechanisms in humans and their neural mechanisms and how these may be applied to improve machine ToM in AI. In addition, we briefly cover some of the limitations still present in creating good AI models with ToM and discuss some ethical implications.

2. *Hot* and *cold* cognition

In recent years researchers have begun to distinguish between *hot* and *cold* cognition (Clark, Chamberlain, & Sahakian, 2009; Lawrence, Clark, Labuzetta, Sahakian, & Vyakarnum, 2008; Roiser & Sahakian, 2013; Roiser, Elliott, & Sahakian, 2012). *Cold* cognition refers to information processing in the absence of any emotional influence. Theoretically, *cold* cognition is engaged on tasks that are emotionally neutral, and includes measures of inhibition, cognitive flexibility, attention and memory. This type of cognition is often used in work or school environments, but also during our daily lives, for example, when you are trying to decide which car to buy. You would likely gather information about different cars, such as their fuel efficiency, safety ratings, and cost, and then analyse this information to determine which car best meets your needs. You might create a pros and cons list to help you compare different options. The process described above is an example of cold decision-making. However, car buying may also have elements of hot, social, and risky decision-making, such as when friends encourage you to spend more money than you feel you could afford on an expensive, flashy sports car that may not be practical for having four passengers or transporting bikes.

Hot cognition specifically engages in tasks that are emotional, or where the outcome is motivationally relevant, and includes measures of emotion recognition, moral judgements and risky decision-making (Roiser & Sahakian, 2013). This type of cognition is particularly used in social interactions. Hot social skills, including theory of mind, are needed for work in the service industry but also for getting along with colleagues and to maximize team performance in the workplace. Hot cognition is also used in motivational or emotional decision-making, for example, when deciding whether to accept a job offer in a different city. While the job offer may be attractive and may offer a better salary and possibly better future career prospects, you may have strong emotional ties to your current city and feel apprehensive about leaving your friends and family. Despite the potential benefits of the new job, you may ultimately decide to remain where you are based on emotional factors, such as a desire to maintain your social connections and avoid feelings of loneliness or isolation. This is an example of hot decision-making, because the decision is influenced by emotional or affective factors rather than a purely rational analysis.

Hot cognition is often associated with impulsive behaviours, whereas *cold* cognition is associated with more thoughtful and deliberate decision-making. For example, when someone is angry or anxious, their *hot* cognition may lead them to react quickly and impulsively, without considering the consequences of their actions. On the other hand, *cold* cognition may be used when making complex decisions, where careful analysis and reasoning are necessary. Both types of cognition are essential for our daily lives, and they work together to help us make decisions that are appropriate for each situation. Understanding the difference between *hot* and *cold* cognition can help us become more aware of how our emotions influence our thoughts and behaviours.

In cognitive neuroscience a number of tests have been developed to measure both *hot* and *cold* cognition. One example, the Cambridge Neuropsychological Test Automated Battery (CANTAB) has several tests to assess, predominantly *cold* cognition including learning and memory, attention, inhibition, executive function and cognitive flexibility, but also some *hot* cognition such as emotion recognition. We provide only a brief description of some tests here, for more detail see (Langley, Sahakian, & Robbins, 2023) or the CANTAB webpage (https://www.cambridgecognition.com/). For example, executive functions are higher-order cognitive processes that include planning and problem-solving. Inhibitory control requires the ability to control our attention, by for example, not responding to a particular stimulus. *Hot* cognition tests may be variants of *cold* cognitive tests that were adapted to include emotionally valanced stimuli or a motivational component. This is the case, for example, of affective go–no go tasks or reinforcement learning tasks. The former, for example, are based upon a well-known *cold* cognitive inhibition task, but because the stimuli being attended to are emotional in nature, this becomes a *hot* cognitive test. EMOTICOM (Bland et al., 2016) is a battery that has been developed to test *hot* cognition specifically. The battery includes measures of emotion recognition, emotional bias, motivation, risky decision-making, and social cognition such as moral judgements and social games. For example, the moral judgement task consists of cartoons of various scenarios: e.g., a cartoon character drops liquid into another character's computer. The subject is asked to judge the amount of shame, guilt, and annoyance and how bad they would feel if in the situation portrayed by the cartoon. The same cartoon is displayed where the subject is the agent or victim, and the action is deliberate or accidental.

In addition, there has been much work to understand the underlying neural mechanisms of these cognitive processes. The neural mechanisms involved in *cold* cognition are primarily located in the prefrontal cortex, the parietal cortex, and the hippocampus (Alexander, DeLong & Strick, 1986). The prefrontal cortex, particularly the dorsolateral prefrontal cortex, plays a crucial role in executive functions, such as working memory, attention, and planning, which are essential for problem-solving and decision-making (Robbins & Everitt, 1996; Otero & Barker, 2013). Frontal cortex is also important for exerting top-down cognitive control over the emotional brain. The parietal cortex is responsible for sensory integration (Macaluso, 2006) and spatial cognition (Burgess, 2008), which allow individuals to perceive and navigate their environment effectively. The hippocampus is involved in episodic memory (de Rover et al., 2011), as well as long-term memory formation and retrieval (Nadel & Moscovitch, 2001), which is essential for learning and knowledge acquisition.

Studies have also shown that the basal ganglia, thalamus, and cerebellum are involved in cold cognition. The basal ganglia are critical for procedural learning and habit formation (Pennartz, Ito, Verschure, Battaglia, & Robbins, 2011; Yin & Knowlton, 2006), while the thalamus is involved in sensory processing and attention (Wimmer et al., 2015). In addition, neurotransmitters, such as dopamine, serotonin, and norepinephrine, also play an essential role in cold cognition. Dopamine is involved in cognitive control, learning, and effort (Westbrook & Braver, 2016), while serotonin and norepinephrine are involved in learning, attention, arousal, and vigilance (Beane & Marrocco, 2004; Coull, 1998; Langley, Armand, et al., 2023). The balance of these neurotransmitters is crucial for optimal cognitive performance.

While many of these regions are also involved in *hot* cognition, there are some regions that are specifically associated with emotion or motivation, such as the ventromedial prefrontal cortex, orbital frontal cortex and amygdala. The prefrontal cortex, particularly the ventromedial prefrontal cortex, is involved in decision-making and emotional regulation (Clark, Cools, & Robbins, 2004; Clark et al., 2008; Hiser & Koenigs, 2018). It helps individuals evaluate and regulate their emotional responses to various stimuli and make appropriate decisions based on the emotional context. The orbital frontal cortex is functionally related to the ventromedial prefrontal cortex and is involved in emotion processing and reward processing in decision making (Kuusinen, Cesnaite, Peräkylä, Ogawa, & Hartikainen, 2018; Rolls, Cheng, & Feng, 2020; Schultz, Tremblay, & Hollerman, 2000) and is often disrupted in depression (Roiser & Sahakian, 2013; Roiser, Rubinsztein, & Sahakian, 2006; Tavares, Drevets, & Sahakian, 2003). The amygdala, located in the limbic system, plays a crucial role in emotional processing, especially fear and anxiety (Dolan & Vuilleumier, 2003; Sergerie, Chochol, & Armony, 2008). It is responsible for detecting and processing emotional stimuli and triggers the body's physiological response to stress.

The hypothalamus and the insula are also involved in *hot* cognition. The hypothalamus is responsible for regulating physiological responses to stress, such as the release of stress hormones and the activation of the sympathetic nervous system (Buijs & Van Eden, 2000; Smith & Vale, 2022). The insula, on the other hand, is involved in interoception, which is the perception of internal bodily states and sensations (Uddin & Menon,

2009; Uddin, Kinnison, Pessoa, & Anderson, 2014). It helps individuals become aware of their emotional states and bodily responses to emotional stimuli. As with cold cognition neurotransmitters also play a key role in *hot* cognition. Dopamine is involved in motivation and reward (Robbins & Everitt, 1992, 1996) and serotonin has been shown to be involved in reward learning, affective and social cognition and moral judgement (Crockett, Clark, Tabibnia, Lieberman, & Robbins, 2008; Kanen, Arntz, Yellowlees, Cardinal, et al., 2021; Kanen, Arntz, Yellowlees, Christmas, et al., 2021; Langley, Armand, et al., 2023).

3. *Hot* and *cold* cognition in theory of mind

Hot and *cold* cognition work together in ToM to help us understand the mental states of others. *Cold* cognition, for example, may help to analyse and interpret social cues, as well as to analyse a person's tone of voice, facial expressions, and body language to infer their mental state. By analysing these cues, we can form a mental model of the speaker's beliefs, desires, and intentions. *Hot* cognition, on the other hand, enables us to empathize with others and regulate our own emotions to gain a deeper understanding of the mental states of others. By combining both of these cognitive processes, we can gain a deeper understanding of others' beliefs, desires, and intentions, can more accurately interpret the mental states of others and respond appropriately in social situations. In addition, it is likely that the best strategy for ToM will be context dependent to some extent, as the best strategy may differ in various situations. The two main strategies for theory of mind are the *theory-theory* approach (Gopnik & Wellman, 1992) and the *simulation-theory* approach (Gordon, 1986). The former can be based on a set of innate rules or on causal and probabilistic reasoning models, whereas the latter is more of a perspective-based approach. Interestingly, the theory-theory approach may be analogous to cold cognition where intellectual processes are used to infer mental states, whereas the simulation-theory approach is more of a hot technique, which relies on one's own motivations and reasoning ability (Gordon, 1986). This is not to suggest that either strategy solely relies on hot or cold cognition, as there is an important interaction between the two (Langley et al., 2022).

Examining individuals with ASD, a neurodevelopmental disorder characterised by dysfunctional ToM, can provide insights into how *hot* and *cold* cognition may work in ToM. Individuals with ASD may actively avoid eye contact, engage in stereotyped or repetitive behaviours and struggle to establish emotional relationships. Interestingly, in high-functioning ASD this is often independent of intelligence (IQ) and other cognitive capacities remain intact (Baron-Cohen, 1995). In addition, it has been suggested that ToM may be independent of other cognitive functions (Langley et al., 2022). One study, (Wheelwright et al., 2006) showed that individuals with ASD showed greater systematising, which is the drive to analyse, understand, predict, control and construct rule-based systems. However, Wheelwright and colleagues found that such individuals showed much lower empathy, which is a specific component of social cognition and is defined as the drive to identify another person's emotions and thoughts, and to respond to these with an appropriate emotion. Systematising is a *cold* cognitive process, while empathising is a *hot* cognitive process. This potentially suggests that *hot* cognition plays a more essential role in ToM.

In terms of the neural mechanisms, neuroimaging research has demonstrated that several brain regions are involved in ToM, including the frontal, temporal and parietal lobes. More specifically, studies have found that the medial prefrontal cortex, anterior cingulate cortex, posterior cingulate cortex, precuneus, temporo-parietal junction, middle temporal gyrus and superior temporal sulcus are all activated in tasks that require ToM (Amodio & Frith, 2006; Saxe, Carey, & Kanwisher, 2004; Schurz, Radua, Aichhorn, Richlan, & Perner, 2014). Interestingly, the results from neuroimaging studies have found an important distinction between cognitive and affective ToM. Cognitive ToM is related to the ability to represent thoughts, intentions and beliefs, whereas affective ToM is more associated with the representation of emotional states and feelings, somewhat like the distinction between *cold* and *hot* cognition. Research has shown that the amygdala, for example, was not involved in tasks of cognitive ToM, but it was significantly involved in affective ToM (Mier et al., 2010; Völlm et al., 2006). One study specifically aimed to examine both the similarities and differences in neural responses between cognitive and affective ToM (Schlaffke et al., 2015). They used the same set of stimuli for all the conditions, but varied the questions asked to prompt either cognitive or affective ToM. The results showed overlapping regions consistent with previous studies of ToM, including the medial prefrontal cortex and temporo-parietal junction but also showed some important differences. The precuneus, cuneus and temporal lobes were recruited in cognitive ToM, whereas the basal ganglia, PCC and prefrontal cortices were involved in affective ToM. In addition, they were able to achieve an 85%

accuracy at distinguishing between the two ToM conditions using a multivariate pattern classifier. This suggests the presence of dissociable neural mechanisms in affective and cognitive ToM, which may reflect the differing influences from *hot* and *cold* cognition.

4. *Hot* and *cold* cognition and theory of mind in AI

AI has made significant strides in recent years, revolutionizing several industries, and changing the way we live and work, a process that continues to evolve. One of the most significant advancements in AI is the development of deep learning (Zhang, Han, & Deng, 2018), which enables machines to process and analyse vast amounts of data and learn from it. This breakthrough has led to significant advancements in natural language processing, image recognition, and speech recognition, enabling machines to perform complex tasks that were previously only possible for humans. Another recent advancement in AI is the development of reinforcement learning (Zhang et al., 2018), which has led to significant breakthroughs in robotics and game-playing. Reinforcement learning is a type of machine learning where machines learn by trial and error, receiving feedback based on their actions. This has enabled machines to learn complex tasks such as playing complex games like chess, go, and poker. Moreover, it has enabled robots to learn and perform complex tasks, such as navigating environments, completing assembly line tasks, and even performing surgical procedures. Most notably, AI has recently made significant strides in the field of natural language processing, allowing machines to understand and generate human–like language. This has led to the development of virtual assistants like Siri, and Alexa which can understand and respond to voice commands. Furthermore, it has enabled machines to analyse and generate written text, leading to the development of AI-powered writing tools that can assist with tasks like content creation, editing, and translation. A paramount example is ChatGPT, a model based on transformer neural networks and generative AI (OpenAI, 2023), which can respond to text prompts and is able to use natural language processing to provide textual responses. In addition, AI has made significant advancements in the field of computer vision (Kakani, Nguyen, Kumar, Kim, & Pasupuleti, 2020; Kitaguchi, Takeshita, Hasegawa, & Ito, 2022), enabling machines to recognize and analyse images and videos. This has led to fast progress in fields like autonomous driving, surveillance, and facial recognition. For instance, self-driving cars use computer vision to detect and analyse their surroundings, enabling them to navigate roads safely and efficiently.

Many of these AI models rely heavily on machine learning algorithms to analyse the data and learn patterns and relationships within it. There are a number of machine learning algorithms, including supervised learning, unsupervised learning and reinforcement learning. Briefly, supervised learning involves training the AI system on labelled data, where the correct output is provided for each input. This allows the system to learn to make predictions or classifications based on new, unlabelled data. Unsupervised learning, on the other hand, involves training the AI system on unlabelled data, allowing it to discover patterns and relationships on its own. This is useful for tasks such as clustering and anomaly detection, for instance. Reinforcement learning, on the other hand, involves training the AI system to make an optimal series of decisions through a process of trial and error, in which it receives feedback based on its actions. This is useful for tasks where there is a reward or penalty associated with each action, such as game-playing or robotics.

AI models are often very successful in limited situations. For example, while AlphaGo is arguably the best Go player in the world, consistently outperforming humans, it would likely not be able to perform well on other games. AlphaZero (McGrath et al., 2022) uses a more general-purpose AI algorithm that can play not just Go, but also chess and shogi (a Japanese board game). Humans, on the other hand, have a more holistic behaviour. Using the comparison with AlphaZero, human chess players have been shown to have better performance on a variety of tasks including memory function (Fattahi, Geshani, Jafari, Jalaie & Mahini, 2015), planning performance (Unterrainer, Kaller, Halsband & Rahm, 2006), creativity (Joseph, Manoharan, Easvaradoss & Chandran, 2017) and even perspective taking (Gao, Chen, Wang & Lin 2019), a component of ToM, which has not been demonstrated in AlphaZero. This demonstrates the need for AI to have a more holistic approach. This becomes of particular importance with the greater interaction between humans and AI.

With the greater reliance in society on AI and its rapid development in domains such as autonomous cars and physical and mental health care, it is essential to ensure the best human-machine interaction. Given that ToM is key to understanding social contexts, this process will undoubtedly require AI models to have social cognition and ToM (Cuzzolin et al., 2020; Langley et al., 2022). For example, autonomous cars are required to interact with human drivers and pedestrians safely. Similarly, in the healthcare setting, to other tasks. As humans we have the ability to deal with multiple, loosely correlated tasks, rather than focusing on a simple, narrow objective. This chapter argued that while there has been significant progress on AI models for specific aspects of ToM, at least in limited settings, there needs to be a more holistic approach. In addition, we highlighted that the more direct use of human data in AI computation has been so far underexplored. This approach might lead to outperform current AI techniques, while allowing us to better benchmark the capabilities of machine ToM. Advancement in this field could be enhanced through interdisciplinary research, including psychologists, neuroscientists, and ethicists and those in mathematics, computing and engineering who have a special interest in AI. This will facilitate enhanced human–machine interaction in healthcare, including in the area of mental health. Importantly, this integrated approach should promote a much-needed ethical framework in AI.

Acknowledgements

Dr Christelle Langley and this research for this article were funded by the Leverhulme Trust, under Research Project Grant RPG-2019–243 to Professor Fabio Cuzzolin and Professor Barbara J Sahakian.

References

Alexander, G. E., DeLong, M. R., & Strick, P. L. (1986). Parallel organization of functionally segregated circuits linking basal ganglia and cortex. *Annual review of neuroscience, 9*(1), 357–381.

Alimardani, M., & Hiraki, K. (2020). Passive brain-computer interfaces for enhanced human-robot interaction. *Frontiers in Robotics and AI, 7.* https://doi.org/10.3389/frobt.2020.00125.

Amodio, D. M., & Frith, C. D. (2006). Meeting of minds: The medial frontal cortex and social cognition. *Nature Reviews. Neuroscience, 7*(4), 268–277. https://doi.org/10.1038/nrn1884.

Baron-Cohen, S. (1995). *Mind-blindnes. An essay opn autism and theory of mind.* London: MIT Press.

Baron-Cohen, S. (1998). Does autism occur more often in families of physicists, engineers, and mathematicians? *Autism: The International Journal of Research and Practice, 2*(3), 296–301.

Baron-Cohen, S., Wheelwright, S., Stott, C., Bolton, P., & Goodyer, I. (1997). Is there a link between engineering and autism? *Autism: The International Journal of Research and Practice, 1*(1), 101–109.

Beane, M., & Marrocco, R. (2004). Norepinephrine and acetylcholine mediation of the components of reflexive attention: Implications for attention deficit disorders. *Progress in Neurobiology, 74*(3), 167–181.

Bland, A. R., Roiser, J. P., Mehta, M. A., Schei, T., Boland, H., Campbell-Meiklejohn, D. K., ... Seara-Cardoso, A. (2016). EMOTICOM: A neuropsychological test battery to evaluate emotion, motivation, impulsivity, and social cognition. *Frontiers in Behavioral Neuroscience, 10*, 25.

Brothers, L. (1990). The social brain: a project for integrating primate behaviour and neurophysiology in a new domain. *Concepts neuroscience, 1*, 27–51.

Buczynski, W., Cuzzolin, F., & Sahakian, B. (2021). A review of machine learning experiments in equity investment decision-making: Why most published research findings do not live up to their promise in real life. *International Journal of Data Science and Analytics, 11*(3), 221–242. https://doi.org/10.1007/s41060-021-00245-5.

Buczynski, W., Steffek, F., Cuzzolin, F., Jamnik, M., & Sahakian, B. (2022). Hard law and soft law regulations of artificial intelligence in investment management. *Cambridge Yearbook of European Legal Studies, 24*, 262–293.

Buijs, R. M., & Van Eden, C. G. (2000). The integration of stress by the hypothalamus, amygdala and prefrontal cortex: Balance between the autonomic nervous system and the neuroendocrine system. *Progress in Brain Research, 126*, 117–132.

Burgess, N. (2008). Spatial cognition and the brain. *Annals of the New York Academy of Sciences, 1124*(1), 77–97.

Christiano, P. (2018). Clarifying "AI Alignment". https://www.alignmentforum.org/posts/ZeE7EKHTFMBs8eMxn/clarifying-ai-alignment.

Clark, L., Cools, R., & Robbins, T. (2004). The neuropsychology of ventral prefrontal cortex: Decision-making and reversal learning. *Brain and Cognition, 55*(1), 41–53.

Clark, L., Chamberlain, S. R., & Sahakian, B. J. (2009). Neurocognitive mechanisms in depression: implications for treatment. *Annual Review of Neuroscience, 32*, 57–74.

Clark, L., Bechara, A., Damasio, H., Aitken, M., Sahakian, B., & Robbins, T. (2008). Differential effects of insular and ventromedial prefrontal cortex lesions on risky decision-making. *Brain, 131*(5), 1311–1322.

Conati, C. (2002). Probabilistic assessment of user's emotions in educational games. *Applied Artificial Intelligence, 16*(7-8), 555–575.

Coull, J. T. (1998). Neural correlates of attention and arousal: insights from electrophysiology, functional neuroimaging and psychopharmacology. *Progress in Neurobiology, 55*(4), 343–361.

Crockett, M. J., Clark, L., Tabibnia, G., Lieberman, M. D., & Robbins, T. W. (2008). Serotonin modulates behavioral reactions to unfairness. *Science (New York, N. Y.), 320*(5884), 1739–1739.

Cuzzolin, F., Morelli, A., Cirstea, B., & Sahakian, B. J. (2020). Knowing me, knowing you: Theory of mind in AI. *Psychological Medicine, 50*(7), 1057–1061.

Dafoe, A. (2018). *AI governance: A research agenda. Governance of AI program.* University of Oxford: Future of Humanity Institute. http://www.fhi.ox.ac.uk/wp-content/uploads/GovAI-Agenda.pdf.

de Rover, M., Pironti, V. A., McCabe, J. A., Acosta-Cabronero, J., Arana, F. S., Morein-Zamir, S., ... Nestor, P. J. (2011). Hippocampal dysfunction in patients with mild cognitive impairment: A functional neuroimaging study of a visuospatial paired associates learning task. *Neuropsychologia, 49*(7), 2060–2070.

Dolan, R. J., & Vuilleumier, P. (2003). Amygdala automaticity in emotional processing. *Annals of the New York Academy of Sciences, 985*(1), 348–355.

Duarte, F. (2023). Number of ChatGPT users (2023). *Exploding Topics.* https://explodingtopics.com/blog/chatgpt-users.

Fattahi, F., Geshani, A., Jafari, Z., Jalaie, S., & Mahini, M. S. (2015). Auditory memory function in expert chess players. *Medical Journal of the Islamic Republic of Iran, 29*, 275.

Future of Life Institute (2023a). Pause giant AI experiments: An open letter https://futureoflife.org/open-letter/pause-giant-ai-experiments/ Accessed 30.10.2023.

Future of Life Institute (2023b). Policymaking in the pause. https://futureoflife.org/wp-content/uploads/2023/04/FLI_Policymaking_In_The_Pause.pdf Accessed 30.10.2023.

Gao, Q., Chen, W., Wang, Z., & Lin, D. (2019). Secret of the masters: Young chess players show advanced visual perspective taking. *Frontiers in Psychology, 10*, 443546.

Gonzalez, B., & Chang, L. J. (2021). *Computational models of mentalizing. The neural basis of mentalizing.* Springer, 299–315.

Gopnik, A., & Wellman, H. M. (1992). Why the child's theory of mind really is a theory. Gordon, R. M. (1986). Folk psychology as simulation. *Mind & Language, 1*(2), 158–171.

Hiser, J., & Koenigs, M. (2018). The multifaceted role of the ventromedial prefrontal cortex in emotion, decision making, social cognition, and psychopathology. *Biological Psychiatry, 83*(8), 638–647.

Ienca, M. (2023). Don't pause giant AI for the wrong reasons. *Nature Machine Intelligence,* 1–2.

Ienca, M., & Vayena, E. (2020). AI ethics guidelines: European and global perspectives. https://rm.coe.int/cahai-2020–07-fin-en-report-ienca-vayena/16809eccac Accessed 30.10.2023.

Jara-Ettinger, J. (2019). Theory of mind as inverse reinforcement learning. *Current Opinion in Behavioral Sciences, 29*, 105–110.

Jin, Z., Levine, S., Gonzalez, F., Kamal, O., Sap, M., Sachan, M., Schölkopf, B. (2022). When to make exceptions: Exploring language models as accounts of human moral judgment. arXiv:2210.01478 https://ui.adsabs.harvard.edu/abs/2022arXiv221001478J Accessed 30.10.2023.

Jobin, A., Ienca, M., & Vayena, E. (2019). The global landscape of AI ethics guidelines. *Nature machine intelligence, 1*(9), 389–399.

Joseph, E., Manoharan, S., Easvaradoss, V., & Chandran, D. (2017). A Study on the Impact of Chess Training on Creativity of Indian School Children. *In CogSci.*

Kakani, V., Nguyen, V. H., Kumar, B. P., Kim, H., & Pasupuleti, V. R. (2020). A critical review on computer vision and artificial intelligence in food industry. *Journal of Agriculture and Food Research, 2*, 100033.

Kanen, J. W., Arntz, F. E., Yellowlees, R., Cardinal, R. N., Price, A., Christmas, D. M., ... Robbins, T. W. (2021). Serotonin depletion amplifies distinct human social emotions as a function of individual differences in personality. *Translational Psychiatry, 11*(1), 81.

Kanen, J. W., Arntz, F. E., Yellowlees, R., Christmas, D. M., Price, A., Apergis-Schoute, A. M., ... Robbins, T. W. (2021). Effect of tryptophan depletion on conditioned threat memory expression: Role of intolerance of uncertainty. *Biological Psychiatry: Cognitive Neuroscience and Neuroimaging, 6*(5), 590–598.

Kim, S. K., Kirchner, E. A., Stefes, A., & Kirchner, F. (2017). Intrinsic interactive reinforcement learning—Using error-related potentials for real world human-robot interaction. *Sci Rep, 7*(1), 17562. https://doi.org/10.1038/s41598-017-17682-7.

Kitaguchi, D., Takeshita, N., Hasegawa, H., & Ito, M. (2022). Artificial intelligence-based computer vision in surgery: Recent advances and future perspectives. *Annals of gastroenterological surgery, 6*(1), 29–36.

Kuusinen, V., Cesnaite, E., Peräkylä, J., Ogawa, K. H., & Hartikainen, K. M. (2018). Orbitofrontal lesion alters brain dynamics of emotion-attention and emotion-cognitive control interaction in humans. *Frontiers in Human Neuroscience, 12.* https://doi.org/10.3389/fnhum.2018.00437.

Langley, C., Sahakian, B., & Robbins, T. (2023). Cambridge neuropsychological test automated battery (CANTAB). In G. Boyle, Y. Stern, D. Stein, B. Sahakian, C. Golden, T.-C. Lee, & A. Chen (Eds.). *The SAGE handbook of clinical neuropsychology.* Clinical Neuropsychological Assessment and Diagnosis.

Langley, C., Cirstea, B. I., Cuzzolin, F., & Sahakian, B. J. (2022). Theory of mind and preference learning at the interface of cognitive science neuroscience, and AI: A review. *Frontiers in Artificial Intelligence, 62.*

Langley, C., Armand, S., Luo, Q., Savulich, G., Segerberg, T., Søndergaard, A., ... Johansen, A. (2023). Chronic escitalopram in healthy volunteers has specific effects on reinforcement sensitivity: A double-blind, placebo-controlled semi-randomised study. *Neuropsychopharmacology: Official Publication of the American College of Neuropsychopharmacology, 48*(4), 664–670.

Lawrence, A., Clark, L., Labuzetta, J. N., Sahakian, B., & Vyakarnum, S. (2008). The innovative brain. *Nature, 456*(7219), 168–169. https://doi.org/10.1038/456168a.

Macaluso, E. (2006). Multisensory processing in sensory-specific cortical areas. *The Neuroscientist, 12*(4), 327–338.

Maréchal, C., Mikołajewski, D., Tyburek, K., Prokopowicz, P., Bougueroua, L., Ancourt, C., & Wegrzyn-Wolska, K. (2019). *Survey on AI-based multimodal methods for emotion detection. High-performance modelling and simulation for big data applications.* Springer, 307–324. https://hal-mines-paristech.archives-ouvertes.fr/hal-02135811.

McGrath, T., Kapishnikov, A., Tomašev, N., Pearce, A., Wattenberg, M., Hassabis, D., ... Kramnik, V. (2022). Acquisition of chess knowledge in alphazero. *Proceedings of the National Academy of Sciences, 119*(47), e2206625119.

Mier, D., Lis, S., Neuthe, K., Sauer, C., Esslinger, C., Gallhofer, B., & Kirsch, P. (2010). The involvement of emotion recognition in affective theory of mind. *Psychophysiology.* https://doi.org/10.1111/j.1469-8986.2010.01031.x.

Nadel, L., & Moscovitch, M. (2001). The hippocampal complex and long-term memory revisited. *Trends in Cognitive Sciences, 5*(6), 228–230.

OpenAI. (2023). ChatGPT. In https://chat.openai.com/chat.

Otero, T. M., & Barker, L. A. (2013). *The frontal lobes and executive functioning. Handbook of executive functioning.* New York, NY: Springer New York 29–44.

Pennartz, C., Ito, R., Verschure, P., Battaglia, F., & Robbins, T. (2011). The hippocampal–striatal axis in learning, prediction and goal-directed behavior. *Trends in Neurosciences, 34*(10), 548–559.

Premack, D., & Woodruff, G. (1978). Does the chimpanzee have a theory of mind? *Behavioral and brain sciences, 1*(4), 515–526.

Robbins, T. W., & Everitt, B. J. (1992). Functions of dopamine in the dorsal and ventral striatum. *Seminars in Neuroscience.*

Robbins, T. W., & Everitt, B. J. (1996). Neurobehavioural mechanisms of reward and motivation. *Current Opinion in Neurobiology, 6*(2), 228–236.

Roiser, J. P., & Sahakian, B. J. (2013). Hot and cold cognition in depression. *CNS Spectrums, 18*(3), 139–149.

Roiser, J. P., Rubinsztein, J. S., & Sahakian, B. J. (2006). Neuropsychology of mood disorders. *Psychiatry, 5*(5), 158–162.

Roiser, J. P., Elliott, R., & Sahakian, B. J. (2012). Cognitive mechanisms of treatment in depression. *Neuropsychopharmacology: Official Publication of the American College of Neuropsychopharmacology, 37*(1), 117–136.

Rolls, E. T., Cheng, W., & Feng, J. (2020). The orbitofrontal cortex: Reward, emotion and depression. *Brain Communications, 2*(2), fcaa196.

Saxe, R., Carey, S., & Kanwisher, N. (2004). Understanding other minds: Linking developmental psychology and functional neuroimaging. *Annual Review of Psychology, 55*(1), 87–124. https://doi.org/10.1146/annurev.psych.55.090902.142044.

Schlaffke, L., Lissek, S., Lenz, M., Juckel, G., Schultz, T., Tegenthoff, M., ... Brüne, M. (2015). Shared and nonshared neural networks of cognitive and affective theory-of-mind: A neuroimaging study using cartoon picture stories. *Human Brain Mapping, 36*(1), 29–39. https://doi.org/10.1002/hbm.22610.

Schultz, W., Tremblay, L., & Hollerman, J. R. (2000). Reward processing in primate orbitofrontal cortex and basal ganglia. *Cerebral Cortex, 10*(3), 272–283.

Schurz, M., Radua, J., Aichhorn, M., Richlan, F., & Perner, J. (2014). Fractionating theory of mind: A meta-analysis of functional brain imaging studies. *Neuroscience & Biobehavioral Reviews, 42*, 9–34. https://doi.org/10.1016/j.neubiorev.2014.01.009.

Sergerie, K., Chochol, C., & Armony, J. L. (2008). The role of the amygdala in emotional processing: A quantitative meta-analysis of functional neuroimaging studies. *Neuroscience & Biobehavioral Reviews, 32*(4), 811–830.

Smith, S. M., & Vale, W. W. (2022). The role of the hypothalamic–pituitary–adrenal axis in neuroendocrine responses to stress. *Dialogues in Clinical Neuroscience.*

Tavares, J. T., Drevets, W., & Sahakian, B. (2003). Cognition in mania and depression. *Psychological Medicine, 33*(6), 959–967.

Taylor, J., & Hern, A. (2023). 'Godfather of AI' Geoffrey Hinton quits Google and warns over dangers of misinformation. *The Guardian*. https://www.theguardian.com/technology/2023/may/02/geoffrey-hinton-godfather-of-ai-quits-google-warns-dangers-of-machine-learning.

Uddin, L. Q., & Menon, V. (2009). The anterior insula in autism: Under-connected and under-examined. *Neuroscience & Biobehavioral Reviews, 33*(8), 1198–1203.

Uddin, L. Q., Kinnison, J., Pessoa, L., & Anderson, M. L. (2014). Beyond the tripartite cognition–emotion–interoception model of the human insular cortex. *Journal of Cognitive Neuroscience, 26*(1), 16–27.

Unterrainer, J. M., Kaller, C. P., Halsband, U., & Rahm, B. (2006). Planning abilities and chess: A comparison of chess and non-chess players on the Tower of London task. *British Journal of Psychology, 97*(3), 299–311.

Völlm, B. A., Taylor, A. N. W., Richardson, P., Corcoran, R., Stirling, J., McKie, S., ... Elliott, R. (2006). Neuronal correlates of theory of mind and empathy: A functional magnetic resonance imaging study in a nonverbal task. *Neuroimage, 29*(1), 90–98. https://doi.org/10.1016/j.neuroimage.2005.07.022.

Westbrook, A., & Braver, T. S. (2016). Dopamine does double duty in motivating cognitive effort. *Neuron, 89*(4), 695–710.

Wheelwright, S., Baron-Cohen, S., Goldenfeld, N., Delaney, J., Fine, D., Smith, R., ... Wakabayashi, A. (2006). Predicting autism spectrum quotient (AQ) from the systemizing quotient-revised (SQ-R) and empathy quotient (EQ). *Brain Research, 1079*(1), 47–56.

Wimmer, R. D., Schmitt, L. I., Davidson, T. J., Nakajima, M., Deisseroth, K., & Halassa, M. M. (2015). Thalamic control of sensory selection in divided attention. *Nature, 526*(7575), 705–709.

Yin, H. H., & Knowlton, B. J. (2006). The role of the basal ganglia in habit formation. *Nature Reviews. Neuroscience, 7*(6), 464–476.

Zeng, Y., Zhao, Y., Zhang, T., Zhao, D., Zhao, F., & Lu, E. (2020). A brain-inspired model of theory of mind. *Frontiers in Neurorobotics, 14*. https://doi.org/10.3389/fnbot.2020.00060.

Zhang, D., Han, X., & Deng, C. (2018). Review on the research and practice of deep learning and reinforcement learning in smart grids. *CSEE Journal of Power and Energy Systems, 4*(3), 362–370.

CHAPTER FIVE

Sense of agency in human-human and human-computer interactions

Sofia Bonicalzi[a,b,*]

[a]Department of Philosophy, Communication and Performing Arts, Roma Tre University, Rome, Italy
[b]CVBE Cognition, Value and Behavior, Ludwig-Maximilians-Universität München, Munich, Germany
[*]Corresponding author. e-mail address: sofia.bonicalzi@uniroma3.it

Contents

1. Introduction	86
2. The sense of agency within the cognitive (neuro)science of action	86
3. Interacting with humans affects the SoA	90
4. Interacting with artificial devices and AI affects the SoA	92
5. Conclusions	97
Funding	98
Conflict of interests	98
References	98

Abstract

The sense of agency (SoA)—a matter that is the subject of lively debate across the philosophy and the cognitive (neuro)science of action—describes the subjective experience and judgement of controlling one's intentional actions and their effects on the outside world. As such, it can be regarded as a fundamental underpinning of moral responsibility for the consequences of one's behaviour. Empirical evidence has shown that the SoA, and consequently our subjective sense of responsibility, tends to be modulated by contextual or external factors such as the fluency of the action selection process and the outcome valence. Crucially, it is also influenced by the presence of other interacting human agents. In addition to these more traditional research topics, recent attention has been directed towards exploring interactions with artificial devices, sometimes perceived as having their own intentionality. This is particularly true for devices that are equipped with forms of artificial intelligence. This perception can also affect the human SoA and responsibility for action. In this chapter, I review this evidence and discuss the conceptual and empirical implications of this line of research.

Developments in Neuroethics and Bioethics, Volume 7
ISSN 2589-2959, https://doi.org/10.1016/bs.dnb.2024.02.006
Copyright © 2024 Elsevier Inc. All rights are reserved, including those for text and data mining, AI training, and similar technologies.

1. Introduction

A key aspect of being a human consists in the first-person experience of being an intentional agent. In the cognitive neuroscience, this sense of agency (henceforth SoA) is defined as the experience of exerting conscious control over actions and their consequences in the outside world (Bonicalzi & Haggard, 2019). Consequently, SoA serves a fundamental biological and cognitive function by making it possible to distinguish *my* actions from actions or events produced by external sources, whether human or non-human (Bayne, 2008). SoA is also a pillar of our moral sense: it grounds first-person responsibility for actions, making us appropriate targets of reactive attitudes (Frith, 2014), and fosters empathy for those who suffer because of our actions (Lepron, Causse, & Farrer, 2015). Empirical evidence has shown that the SoA, and the associated sense of responsibility, are modulated by both physiological and contextual factors (Christensen, Di Costa, Beck, & Haggard, 2019; Herman & Tsakiris, 2020; Limerick, Coyle, & Moore, 2014; Sato & Yasuda, 2005; Wenke, Fleming, & Haggard, 2010), as well as, crucially, by social variables (Beyer, Sidarus, Bonicalzi, & Haggard, 2017; Caspar, Christensen, Cleeremans, & Haggard, 2016; El Zein, Dolan, & Bahrami, 2022). Besides such traditional research topics, however, attention has been recently paid to how interacting with artificial devices, some of which—and especially those that are equipped with forms of artificial intelligence (AI)—are possibly perceived as having an intentionality of their own, may also affect the SoA (Barlas, 2019; Ciardo, Beyer, De Tommaso, & Wykowska, 2020; Lepron et al., 2015; Roselli, Ciardo, De Tommaso, & Wykowska, 2022).

In the chapter, I will first provide a brief overview of some key problems in the literature on the SoA (§ 2). Then, I will discuss how the social dimension, especially in collaborative tasks, may modulate the SoA (§ 3) and, finally, how interacting with artificial devices, and notably with AI, may also impact the SoA (§ 4).

2. The sense of agency within the cognitive (neuro) science of action

Over the last two decades, the philosophy and the cognitive neuroscience focusing on mind and action have shown a profound interest in the SoA, with the aim of constructing theoretical and empirical models to understand its origin and function within our mental life. In this regard, Synofzik, Vosgerau, and Newen (2008) proposed a now *standard* distinction between the *feeling of agency* (FoA) and the *judgement of agency* (JoA). Pre-conceptual and sub-personal, the FoA is rooted in sensorimotor processes and involves only an implicit representation of the agent's self. Conversely, the JoA represents the conceptual, interpretative, and conscious component of the SoA. The FoA may or may not be followed by the JoA: the former is indeed often perceived just as a coherent flow, without the subject being fully aware of the control exerted over actions.

Two computational accounts, traditionally seen as mutually exclusive, have shaped the debate on the origin of the SoA. The predictive-prospective models, the most well-known being the *comparator model* (Blakemore, Wolpert, & Frith, 2002), give centrality to the agent's intentions and explain the SoA as a direct consequence of the processes occurring within the motor system when the agent forms the intention to act: the agent can anticipate, albeit at a sub-personal and non-conceptual level, the outcomes of her intentional actions, that is, of the actions that are caused by her intentions. The comparator model is thus based on multiple comparison mechanisms that match different representations and predictions of the agent's state. The SoA, signalling the match between expectations and actual outcomes, would then arise as the experiential trace of the motor control exerted over action selection and planning.

Conversely, retrospective-inferential models (Wegner, 2002) rest on an epiphenomenalist conception of the mental life, according to which mental states are epiphenomena, with no causal effect on the behaviour or functioning of the body. As such, they emphasise the role of information and processes external to the motor system, which depend on the environmental and social context in which the action takes shape. Since humans do not exercise real conscious control over actions, which are caused by unconscious neural mechanisms, the SoA must be illusory as well and derived from retrospective confabulatory practices.

In recent years, convincing evidence has led to integrating the two accounts in what is known as *cue-integration theory*, according to which both predictive and postdictive cues—that is, different sensory modalities and cognitive processes—would contribute, to varying degrees depending on the situation, to the overall SoA. In humans, and presumably in other species, the integration between the various signals could take place according to a Bayesian process, by which the sources of information and available evidence are integrated to obtain an estimate or decide about an event (Pacherie, 2014). In the case of the SoA, the result of the integration depends on the strength of the predictive priors and the weight of the available evidence, which in turn is determined by their degree of contextual reliability. For instance, when the agent is insecure about having caused an effect, the priors will be weak. However, a strong SoA may still result from the sensorimotor or perceptual feedback, if assessed as reliable. Conversely, strong priors can compensate for weak sensorimotor or perceptual feedback. If, on the other hand, both signals are weak, the SoA will itself be weak, or the action could be attributed to an external source altogether (Fletcher & Frith, 2009).

A key issue in the SoA literature is, however, how to exactly measure this phenomenon. The SoA is notoriously a "thin" and "evasive" signal (Metzinger, 2006), about which people tend to have minimal awareness (Moore, 2016). The available measures can be divided into two camps, that is, they can be explicit or implicit. Explicit measures are based on subjects' reports about past performance, with subjects indicating the extent to which they have experienced a SoA by using a binary or a continuous visual analogue scale ranging from a low level to a high level of control (Pacherie, 2014). Explicit measures are criticised for various reasons. One challenge is that participants are not usually accustomed to reporting something as elusive as the SoA. The risk is that they misrepresent the task demands, replacing them with more familiar ones, for example, by providing a metacognitive assessment of how sure they are of having caused an event (Grünbaum & Christensen, 2020). Moreover, it is well known that explicit measures are particularly susceptible to biases, including self-serving biases to preserve self-esteem in case of negative outcomes (El Zein et al., 2022).

Given these and other critical issues, several scientists have turned to implicit measures or combinations of implicit and explicit measures. Implicit measures treat behavioural or neurophysiological parameters related to action performance as proxies of the SoA. These measures are said to be "implicit", as participants are not asked to directly assess their SoA. Compared to explicit measures, they are believed to be less prone to biases and illusions and, therefore, more reliable. Two of the most widely used continuous implicit measures are *sensory attenuation in self-touch* and the *intentional binding effect* (IB). Sensory attenuation, especially in the tactile, visual, and auditory domains, serves as a proxy of the SoA to the extent that self-produced actions cause attenuated bodily sensations when compared with movements produced externally. In accordance with the comparator model, this effect might depend on the easily available predictions allowed by self-produced actions, which make the extra signals generated by the sensory receptors redundant (Haggard, 2017).

The IB is based on mental chronometry, a methodology used in cognitive science to measure the duration and temporal order of mental processes. The basic idea underlying the IB is that intentional actions modify our temporal perception of the action and the effect, producing a sort of temporal attraction between them. In particular, (i) the intentional action producing an effect is subjectively perceived as forward in time (compared to a baseline condition where the intentional action produces no effect) and (ii) the effect of an intentional action is subjectively perceived as back in time (compared to a baseline condition where the effect is produced in the absence of an intentional action). Furthermore, the subjectively perceived temporal interval between the intentional action (usually a keypress) and the effect (usually a sound) is shorter than the interval we observe (when the action is involuntary) (Haggard, Clark, & Kalogeras, 2002). According to its proponents, and on the grounds of the comparator model, this temporal attraction between the action and the effect is determined by the same motor mechanisms that are responsible for allowing the predictions of the sensory consequences of intentional actions. By bringing the action and the effect closer in time, the IB would then reinforce the association between the intention and the outcome of the action.

On the one hand, the claim that explicit and implicit measures (especially the IB) refer to the same underlying construct (if not exactly to the same phenomenon) is supported by the observation that various variables affect both in comparable manners. For instance, both implicit and explicit measures are boosted by the presence of motivations to act (Borhani, Beck, & Haggard, 2017), by the positivity of the outcome (Beyer et al., 2017), by having multiple options to choose from (Barlas et al., 2018), or by the fluency of the action selection process (Sidarus, Vuorre, & Haggard, 2017). On the other hand, based also on some notable mismatches between explicit and implicit measurements, alternative explanations suggest that the IB be not a signature of intentional action, but the effect of grasping the causal connection between the action and the effect (Buehner, 2012) so that its prominence within the debate on the SoA ought to be downsized.

Other implicit measures of the SoA consist in neural markers investigated with imaging methods, such as electroencephalography (Beyer et al., 2017) or magnetoencephalography (El Zein et al., 2022). An EEG-based example is the *feedback-related negativity* (FRN), an event-related potential—that is, a brain response to a sensory, cognitive, or motor stimulus—occurring about 220–350 ms after outcome presentation. The amplitude or the peak (about 250 ms after outcome presentation) of the FRN is taken as revealing information regarding how the agent subjectively processed the action: the FRN tends to be larger for negative outcomes and varies depending, among other factors, on the perceived controllability of the outcomes and the level of monitoring of the consequences of the action (San Martín, 2012).

As mentioned, multiple variables may impact the SoA, sometimes leading to errors and illusions of control, where the actual sensorimotor control does not match participants' experience. For instance, a heightened or weakened SoA may be induced through the usage of subliminal *primes* (Wenke et al., 2010) or appropriately distorted *feedback* (Limerick et al., 2014), as well as modulated by outcome predictability, with more predictable outcomes linked with an increase in the SoA (Sato & Yasuda, 2005). In addition to contextual variables, the agent's physiological states, including interoceptive (Herman & Tsakiris, 2020) and emotional states (Christensen et al., 2019), can also influence the SoA. As we will see in the next section, a particularly important variable, deeply affecting the SoA, is the social context in which the action takes place.

3. Interacting with humans affects the SoA

In social situations, the activation of specific socio-cognitive processes and networks allows people to draw inferences regarding the thoughts and intentions of their peers (Wiese, Metta, & Wykowska, 2017), fostering the tendency to imitate them (Frith, 2014) and to engage in prosocial behaviours (Pacherie, 2014). In this context, the ability to develop a SoA is crucial both for distinguishing *my* action from the action of another, but also for co-planning and sharing experiences, goals, and plans with others (Antusch, Custers, Marien, & Aarts, 2020). An open question is thus how the social context modulates the individual SoA. However, as this section aims to make clear, no unified answer is available—partly because, at the end of the day, not all social contexts share the same features.

Social psychology has indeed shown for decades that the social context, including when other people are merely present but inactive, may have multifarious (and sometimes pernicious) effects on action processing. Classic studies show how, when outcomes are positive, individuals overestimate their SoA, attributing the group's success to their merits (Farwell & Wohlwend-Lloyd, 1998). Conversely, when outcomes are negative, people deflect responsibility and blame onto the group. This phenomenon is known as "diffusion of responsibility" and is linked to the infamous "bystander effect"—the tendency to avoid helping individuals in distress when others are present and may also intervene (Darley & Latané, 1968). Other studies have shown that individuals feel less regretful when negative outcomes result from a group activity, even when one's decision would not have deviated from the majority (Nicolle, Fleming, Bach, Driver, & Dolan, 2011). And other well-known studies, such as (Milgram's on coercion 1963) and Bandura's on aggressive tendencies in groups (1991), show how the social context may decrease people's sense of responsibility, stress, tension, and self-blame.

In many social situations, reductions of the SoA might be due to self-serving biases, reflected in explicit control ratings, put forward to preserve self-esteem. However, recent evidence shows that such a decrease in the SoA in social contexts corresponds to measurable changes in implicit measures of agency as well (including the already mentioned FRN), suggesting that subjects process the action differently when other people are present. In this case, a possible explanation is that in social contexts, the agent must engage in complex mentalisation processes to interpret the thoughts of others and plan accordingly. These mentalisation processes would reduce the *fluency* of the action selection process, resulting in an increase in reaction time and a decrease in the SoA (Beyer et al., 2017; El Zein et al., 2022).

In more cooperative situations, people may rather engage in what the relevant literature has baptised as "joint actions", that is, actions in which people act together in view of a common goal (Zapparoli, Paulesu, Mariano, Ravani, & Sacheli, 2022). It is debated whether, in joint actions, people must necessarily have a clear mutual knowledge of the common goal to favour the sharing and alignment of their mental states, that is, their intentions, knowledge, and perspectives (Gallotti, Fairhurst, & Frith, 2017). In any case, such situations require people to both control their actions and to interpret and monitor what the co-actors are doing, mentally integrating their reciprocal contributions, and combining—along the lines of the cue integration theory—a multiplicity of prospective and retrospective signals.

The effect of joint actions on the SoA can be considered based on parameters that are not intrinsically social (e.g., predictability of the outcome, familiarity with the task, fluency in the action selection process), or on parameters that are more eminently social, such as the level of coordination achieved or the specific role the agent plays, for example, follower or leader. Focusing on the latter, research has shown that reductions in the individual SoA may occur in situations where the agent is prevented from monitoring what is going on. This typically happens, for instance, in forms of hierarchical or asymmetrical cooperation, where the agent occupies a subordinate position and cannot contribute to the overall action planning. In such cases, the SoA may be determined more by the agent's ability and opportunity to predict the outcomes rather than by the actual (sensorimotor) control. Conversely, those at the top of the hierarchy, with extended capacities and opportunities to plan and monitor the action, may experience a heightened SoA. While social situations may often promote a decrease in the individual SoA, in some cases an enhancement in the SoA can be explained in the light of the broadening of action possibilities offered by acting jointly with others. The key notion here is that of "we agency", accompanied by a collective SoA in situations in which the contribution of the co-players tends to be egalitarian, less differentiated, and more symmetrical (Pacherie, 2014; Zapparoli et al., 2022).

4. Interacting with artificial devices and AI affects the SoA

As mentioned, in the case of human-human interactions, modulations to the individual SoA might be significantly due to the fluency-disrupting mentalization efforts that are required to attribute mental states to others and monitor their behaviour (Beyer et al., 2017; El Zein et al., 2022). In today's rapidly evolving technological landscape, however, an increasing number of interactions involve AI and artificial agents. While fluid human-computer interactions seemingly require the co-operators to construct models of the (shared) action plans, it is not straightforward whether the said mentalization processes occur in such contexts.

On the one hand, it has been observed that the attribution of intentionality to artificial agents—possibly fuelled by the human-like or intelligent features that some of them display—enhances human-computer interactions, in terms of increased connectedness, empathy, and pro-sociality, which in turn can improve cooperation (Wiese et al., 2017). In this regard, humans may even experience emotion-based attitudes towards machines, including trust or aversion. Trust is facilitated by forms of transparent AI, both in terms of explainability and in terms of the ability of the system to share confidence and performance parameters, usually directed towards other humans (Zerilli, Bhatt, & Weller, 2022). On the other hand, attributing intentionality to machines does not always occur and may be influenced by minor contextual factors. Overall, the mere fact that artificial agents share our same physical space does not necessarily imply that they are located in our same cognitive and social dimension (Ziemke, 2023).

Some pieces of evidence seemingly support the claim that, indeed, ascribing intentional states to partner-like artificial agents may depend, at least in part, on whether they are perceived as similar to us, that is, endowed with a mind, capable of having experiences, feeling emotions, or developing prosocial or competitive attitudes: collaborative, or even shared, actions might be then facilitated when artificial agents have a human appearance (Limerick et al., 2014), or features (e.g., the tone of voice) prompting certain responses on the human side (Reeves & Nass, 1996). At the same time, the well-known phenomenon of the "uncanny valley" refers to a concept used in psychology and robotics to describe the disquiet we feel when interacting with realistic representations of human beings, but which also have obvious artificial characteristics. The phenomenon stems from a categorisation conflict: the initially positive attitudes towards them become progressively more pronounced the more the artificial agents are perceived as similar to us, but decay drastically when we realise that they are not exactly human. Overcoming discomfort could be made possible by building artificial agents with such a level of realism that they are not perceived as disturbing (Wiese et al., 2017).

Key questions for current and future research concern, therefore, how interacting with AI and artificial agents may critically affect the SoA, for example, by causing mentalization efforts analogous to the ones that are in place in human–human interactions. Importantly, for the former to be the case, the artificial agent must be perceived as a partner, and not merely as a tool. When artificial devices work as tools, it is indeed known from previous research that they may affect the SoA in specific manners. This is part of an ongoing discussion regarding the impact on human agency that has been associated with the diffusion of AI-based algorithms and, more generally, artificial devices (Bonicalzi, De Caro, & Giovanola, 2023).

However, even in the absence of fluency-disrupting mentalization efforts due to intentionality attributions, variations in the SoA as related to the relationship with artificial systems of various sorts may be traced back to the mechanisms of the comparator model or the cue integration theory, whereby the interference of the artificial agent increases or decreases the possibility of making accurate predictions about the congruence between the current and the future state of the system (Coyle, Moore, Kristensson, Blackwell, & Fletcher, 2012; Limerick et al., 2014). For instance, in the computer-based task by Coyle et al. (2012), the IB is enhanced when the normal keyboard is replaced by the so-called "skinput", a tool that allows users to interact with other electronic devices using their skin as an interface. Compared to more conventional tools, the skinput either responds better to commands, supports a higher degree of congruence between predicted and actual sensory outcomes, or even provides additional sensory evidence, which in turn impacts the SoA.

To empirically test differences in attitudes towards multiple types of artificial agents, Ciardo et al. (2020) adapt a diffusion of responsibility task by Beyer et al. (2017). The original experiment included two conditions: *individual* and *social*. In the individual condition, a human subject had to stop a marble rolling down a bar at a variable speed. In the social condition, two human participants are tricked into believing that they must work cooperatively to achieve the same goal (unbeknownst to them, they are, however, both playing with a computer). However, for the marble to be stopped, it is sufficient that one of the co-player acts. As acting is costly, both co-players are incentivised to avoid acting, waiting for the other's intervention. However, the task is programmed so that the partner (i.e., the computer) acts rarely, forcing the participant to intervene. The results show that the SoA is reduced in the social condition, as said possibly due to the mentalization efforts needed to monitor the co-player's intentions under uncertainty, which in turn decreases the perceived control over action—and this even when the co-player is present but inactive, *de facto* without the participant's sensorimotor control over the action being altered with respect to the individual condition.

Ciardo et al. (2020) modify the original task, by adding two extra conditions in which humans must cooperate with a robot with humanoid characteristics (Cozmo, a domestic AI robot with the appearance of a small truck and equipped with light screens with which it expresses emotions and reactions) or with a non-humanoid mechanical device. The results show that the SoA is similarly reduced in the human–human and the human–Cozmo interactions, but not when humans interact with the non-humanoid device. The difference might be interpreted as suggesting that it is the effort in decoding Cozmo's *mental states*—and not simply a reduction in the agent's predictive accuracy due to the presence of an extra actor—that causes a reduction in the agent's SoA.

In a follow-up study, Roselli and colleagues (2022) make humans interact with the iCub—a humanoid AI robot designed for research in cognitive robotics. The iCub looks like a three-year-old child, has an anthropomorphic body, mobile arms and hands, and a head equipped with stereo cameras and sensors. The study shows that participants can develop a sort of *vicarious* SoA for the actions performed by the iCub and that the intensity of this vicarious SoA can be predicted based on the level of intentionality attributions.

A different line of research explores variations in the SoA in social situations where the agent is imposed an action by an artificial co-player. It is known from previous research that when participants obey the experimenter's orders, their SoA tends to decrease. For instance, in Caspar et al. (2016), which drew inspiration from Milgram's (in)famous experiment on coercion, participants are instructed by the human experimenter to deliver electric shocks to a co-player, in exchange for a small amount of money. The results, based on both implicit and explicit measures, show that participants experience less SoA in this condition as compared to when they are free to deliver the shocks. A fMRI-based follow-up (Caspar, Beyer, Cleeremans, & Haggard, 2021) suggests that when they simply obey orders, people feel less empathetic and guilty regarding the pain of others, even when they are personally responsible for causing it. A possible interpretation is that, in such contexts, participants transfer their SoA and responsibility to the experimenter, representing themselves as a mere link within a wider causal chain, in this case originating from the experimenter's command.

A key question is thus whether artificial agents can be perceived as entities able to originate *new* causal chains and thus be attributed responsibility in place of the participant in case of imposed actions. In Lepron et al. (2015), subjects are led to believe that they have been selected to participate in a clinical trial designed to test a painful medical treatment. In one of the proposed experiments, they are either directly responsible for administering the treatment or participate as observers. When participants are directly responsible for inflicting pain, their affective empathic response to the patients' painful reaction increases. In another experiment in the study, a manipulation prevents participants from clearly understanding who has initiated the action. The results show that the more participants feel responsible for inflicting pain (so, the more SoA they experience), the greater they perceive a sense of unpleasantness for the situation. This indicates that the SoA influences the way we perceive the reactions, intentions, and emotions of others, playing a sort of regulatory role in guiding our moral conduct and supporting our ability to experience "reactive attitudes" (Strawson, 1962) *appropriate* to the context, including empathy towards the sufferers. This contributes to explaining why SoA modulations, possibly acting upon our action–related reactive attitudes, are a delicate matter. In this respect, in another condition of Lepron et al. (2015), participants administer the medical treatment following the computer's instructions. In this case, the empathic response is similar to the one observed when participants spontaneously administer the treatment. This seems to indicate that, in this case, the artificial agent (the computer) is not perceived as a real intentional actor onto whom agency and responsibility can be transferred.

Conversely, Barlas (2019) indicates how both human partners and humanoid robots could influence the agent's SoA in situations of imposed actions. In the study, both the IB and the explicit control ratings drop when participants are told what to do by their human or artificial partner. In other cases, the interaction with artificial agents is such that it even cancels out the human SoA. In a study with the IB, Obhi and Hall (2011) propose instead that participants engage in a series of shared actions, with a human or a computer as partners. The IB arises only in the former case, while being absent in the latter. Such findings give rise to potential concerns about the possibility of developing artificial agents that collaborate with humans without nullifying their SoA and responsibility.

As this brief review aimed to convey, it is not an easy task to piece together all these, somehow contradictory, results. As said, this discussion reflects a broader discourse concerning how human reasoning and action are influenced by AI and robotics. It is indeed already known that the failure to exercise increasingly outsourced skills, as caused by the interaction with AI and artificial devices, can contribute to skill degradation, to the uncritical and passive adherence to the responses offered by artificial agents that appear more performant and competent (*automation complacency*), and to the preference for automated solutions deemed as more reliable (*automation bias*) (Bonicalzi, 2023).

Regarding, more specifically, the SoA, it can be provisionally suggested that the various ways in which human–computer interactions take place, as well as the features of humanoid or non-humanoid artificial agents, may have a variable impact on the human SoA. While several studies seem to suggest that such interactions bring about a reduction in the SoA, there is also evidence that when computer assistance is not pervasive and leaves room for human initiative, the SoA can be at least partially preserved. In some cases, collaborating with artificial agents, be they considered as tools or as co-players, may even lead to an expansion in people's agentive possibilities, in terms of their ability to control wider environments, real and virtual, beyond the immediate context of the action (Haggard, 2017).

It is thus no coincidence that in the literature on user experience, it has been often emphasised that human-computer-interactions and brain-computer-interfaces should be designed in such a way that they support the internal locus of control, by preserving the agent's sense of controlling the situation in which they are immersed and of relying on a system that predictably responds to her commands and actions (Moore, 2016). This requires carefully calibrating the way the input is fed into the system, the reliability with which the output is produced, the type of feedback produced by the system, and the interval between actions and consequences (Limerick et al., 2014). Furthermore, it has been suggested that implementing forms of explainable AI (albeit with notorious feasibility problems) into artificial devices—which permits humans to develop a better understanding of how the artificial co-agent makes decisions and plans actions—may help re-establish the SoA of the human operators, for instance by boosting their confidence and their level of acceptance of artificial agents (Pagliari, Chambon, & Berberian, 2022).

The problem of how to refine interactions with artificial agents is also a key issue in social robotics, an interdisciplinary research field that combines robotics, AI, and social sciences to develop robots designed to interact and collaborate with humans in a fluid and intuitive way. In this regard, embodied and situated theories of AI applied to robotics emphasise that the physical basis and sensorimotor apparatuses of modern AI systems make it possible to overcome the limitations of classical computational approaches, according to which AI cannot be endowed with a form of intentionality. Instead, the new AI systems might be more and more equipped with artificial forms of intentionality, at least understood as the ability to express intelligible intentions and act with a purpose (Ziemke, 2023) as well as to produce appropriate affective or cognitive responses, developing an ability to exhibit human-like empathic attitudes (Wiese et al., 2017). In recent years, there has been a focus on how to implement the SoA in artificial intelligent agents. This may eventually permit AI to develop a subjective understanding of its own agency, possibly achieving better results in terms of both action control and sense of social responsibility (Legaspi, He, & Toyoizumi, 2019).

5. Conclusions

The applications of AI and robotics, from interactive learning to healthcare, are manifold and raise a wide array of technological, juridical, and ethical issues. These range from the problem of safeguarding users' privacy to the impact that these technologies may have on human rela-tionships. In addition, as this chapter aimed to emphasise, cognitive (neuro) sciences have highlighted how the interaction with AI and other automated systems to support daily and professional activities brings with it alterations to human cognition and action, affecting our SoA—defined as the sub-jective experience and judgement of controlling one's intentional actions and their effect in the outside world. As these systems have emerged as transformative and pervasive forces, with profound and appreciable implications for various industries and aspects of our daily lives, carefully balancing these considerations is crucial when building AI systems and robots that can cooperate with humans fruitfully and harmoniously.

Funding

The author benefitted from the PRIN Grant 20175YZ855 - New challenges for applied ethics. The moral impact of scientific and technological advances, from the Italian Ministry for Education, University, and Research (Ministero dell'Istruzione, dell'Università e della Ricerca).

Conflict of interests

The author has no conflicts of interest to declare.

References

Antusch, S., Custers, R., Marien, H., & Aarts, H. (2020). Intentionality and temporal binding: Do causality beliefs increase the perceived temporal attraction between events? *Consciousness and Cognition, 77*, 102835.

Bandura, A. (1991). Social cognitive theory of self-regulation. *Organizational Behavior and Human Decision Processes, 50*(2), 248–287.

Barlas, Z. (2019). When robots tell you what to do: Sense of agency in human- and robot-guided actions. *Consciousness and Cognition, 75*, 102819.

Bayne, T. 2008, The phenomenology of agency, in "Philosophy Compass" 3, 11, pp. 182–202.

Blakemore, S. J., Wolpert, D. M., & Frith, C. D. (2002). Abnormalities in the awareness of action. *Trends in Cognitive Sciences, 6*(6), 237–242.

Beyer, F., Sidarus, N., Bonicalzi, S., & Haggard, P. (2017). Beyond self-serving bias: Diffusion of responsibility reduces sense of agency and outcome monitoring. *Social Cognitive and Affective Neuroscience, 11*(12), 138–145.

Bonicalzi, S. (2023). A matter of justice. The opacity of algorithmic decision-making and the trade-off between uniformity and discretion in legal applications of artificial intelligence. *Teoria. Rivista Di Filosofia, 42*(2), 131–147.

Bonicalzi, S., De Caro, M., & Giovanola, B. (2023). Artificial intelligence and autonomy: On the ethical dimension of recommender systems. *Topoi, 42*, 819–832.

Bonicalzi, S., & Haggard, P. (2019). From freedom from to freedom to: New perspectives on intentional action. *Frontiers in Psychology, 10*, 1193.

Borhani, K., Beck, B., & Haggard, P. (2017). Choosing, doing, and controlling: Implicit sense of agency over somatosensory events. *Psychological Science, 28*(7), 882–893.

Buehner, M. J. (2012). Understanding the past, predicting the future: Causation, not intentional action, is the root of temporal binding. *Psychological Science, 23*(12), 1490–1497.

Caspar, E. A., Beyer, F., Cleeremans, A., & Haggard, P. (2021). The obedient mind and the volitional brain: A neural basis for preserved sense of agency and sense of responsibility under coercion. *PLoS One, 16*(10), e0258884.

Caspar, E. A., Christensen, J. F., Cleeremans, A., & Haggard, P. (2016). Coercion changes the sense of agency in the human brain. *Current Biology: CB, 26*(5), 585–592.

Christensen, J. F., Di Costa, S., Beck, B., & Haggard, P. (2019). I just lost it! Fear and anger reduce the sense of agency: A study using intentional binding. *Experimental Brain Research, 237*(5), 1205–1212.

Ciardo, F., Beyer, F., De Tommaso, D., & Wykowska, A. (2020). Attribution of intentional agency towards robots reduces one's own sense of agency. *Cognition, 194*, 104109.

Coyle, D., Moore, J., Kristensson, P. O., Blackwell, A. F., & Fletcher, P. C. (2012). I did that! Measuring users' experience of agency in their own actions. In *CHI, ACM conference on human factors in computing systems*, 2025–2034.

Darley, J. M., & Latane, B. (1968). Bystander intervention in emergencies: Diffusion of responsibility. *Journal of Personality and Social Psychology, 8*, 377–383.

El Zein, M., Dolan, R. J., & Bahrami, B. (2022). Shared responsibility decreases the sense of agency in the human brain. *Journal of Cognitive Neuroscience, 34*(11), 2065–2081.

Farwell, L., & Wohlwend-Lloyd, R. (1998). Narcissistic processes: Optimistic expectations, favorable self-evaluations, and self-enhancing attributions. *Journal of Personality, 66*(1), 65–83.

Fletcher, P. C., & Frith, C. D. (2009). Perceiving is believing: A Bayesian approach to explaining the positive symptoms of schizophrenia. *Nature Neuroscience, 16*, 48–58.

Frith, C. D. (2014). Action, agency and responsibility. *Neuropsychologia, 55*, 137–142.

Gallotti, M., Fairhurst, M. T., & Frith, C. D. (2017). Alignment in social interactions. *Consciousness and Cognition, 48*, 253–261.

Grünbaum, T., & Christensen, M. S. (2020). Measures of agency. *Neuroscience of Consciousness, 2020*(1), niaa019.

Haggard, P. (2017). Sense of agency in the human brain. *Nature Reviews: Neuroscience, 18*, 196–207.

Haggard, P., Clark, S., & Kalogeras, J. (2002). Voluntary action and conscious awareness. *Nature Neuroscience, 5*, 382–385.

Herman, A. M., & Tsakiris, M. (2020). Feeling in control: The role of cardiac timing in the sense of agency. *Affect Science, 1*(3), 155–171.

Legaspi, R., He, Z., & Toyoizumi, T. (2019). Synthetic agency: Sense of agency in artificial Intelligence. *Current Opinion in Behavioral Sciences, 29*, 84–90.

Lepron, E., Causse, M., & Farrer, C. (2015). Responsibility and the sense of agency enhance empathy for pain. *Proceedings of the Royal Society B, 282*(1828), 20142288.

Limerick, H., Coyle, D., & Moore, J. W. (2014). The experience of agency in human-computer interactions: A review. *Frontiers in Human Neuroscience, 8*, 643.

Metzinger, T. (2006). Conscious volition and mental representation: Toward a more fine-grained analysis. In N. Sebanz, & W. Prinz (Eds.). *Disorders of volition*. MIT Press.

Milgram, S. (1963). Behavioral study of obedience (PDF). *Journal of Abnormal and Social Psychology, 67*(4), 371–378.

Nicolle, A., Fleming, S. M., Bach, D. R., Driver, J., & Dolan, R. J. (2011). A regret-induced status quo bias. *Journal of Neuroscience, 31*(9), 3320–3327.

Obhi, S. S., & Hall, P. (2011). Sense of agency and intentional binding in joint action. *Experimental Brain Research, 211*(3), 655.

Pacherie, E. (2014). How does it feel to act together? *Phenomenology and the Cognitive Sciences, 13*, 25–46.

Pagliari, M., Chambon, V., & Berberian, B. (2022). What is new with artificial intelligence? Human–agent interactions through the lens of social agency. *Frontiers in Psychology, 1*.

Reeves, B., & Nass, C. (1996). *The media equation: How people treat computers, television and new media like real people and places*. New York: Cambridge University Press.

Roselli, C., Ciardo, F., De Tommaso, D., & Wykowska, A. (2022). Human-likeness and attribution of intentionality predict vicarious sense of agency over humanoid robot actions. *Scientific Reports, 12*(1), 13845.

San Martín, R. (2012). Event-related potential studies of outcome processing and feedback-guided learning. *Frontiers in Human Neuroscience, 6*, 304.

Sato, A., & Yasuda, A. (2005). Illusion of sense of self-agency: Discrepancy between the predicted and actual sensory consequences of actions modulates the sense of self-agency, but not the sense of self-ownership. *Cognition, 94*(3), 241–255.

Sidarus, N., Vuorre, M., & Haggard, P. (2017). How action selection influences the sense of agency: An ERP study. *Neuroimage, 150*, 1–13.

Strawson, P. (1962). Freedom and resentment. *Proceedings of the British Academy, 48*, 187–211.

Synofzik, M., Vosgerau, G., & Newen, A. (2008). Beyond the comparator model: A multifactorial two-step account of agency. *Consciousness and Cognition, 17*(1), 219–239.

Wegner, D. M. (2022). *The illusion of conscious will*. Cambridge, MA: MIT Press.

Wenke, D., Fleming, S. M., & Haggard, P. (2010). Subliminal priming of actions influences sense of control over effects of action. *Cognition, 115*(1), 26–38.

Wiese, E., Metta, G., & Wykowska, A. (2017). Robots as intentional agents: Using neuroscientific methods to make robots appear more social. *Frontiers in Psychology*.

Zapparoli, L., Paulesu, E., Mariano, M., Ravani, A., & Sacheli, L. M. (2022). The sense of agency in joint actions: A theory-driven meta-analysis. *Cortex; A Journal Devoted to the Study of the Nervous System and Behavior, 148*, 99–120.

Zerilli, J., Bhatt, U., & Weller, A. (2022). How transparency modulates trust in artificial intelligence. *Patterns, 3*(4) Article 100455.

Ziemke, T. (2023). Understanding social robots: Attribution of intentional agency to artificial and biological bodies. *Artificial Life, 29*(3), 351–366.

CHAPTER SIX

Anthropomorphism in social AIs: Some challenges

Arleen Salles[a,*] and Abel Wajnerman Paz[b]

[a]Institute of Neuroethics (IoNx), Think and Do Tank, Atlanta, Georgia, USA
[b]Pontificial Universidad Catolica de Chile, Instituto de Eticas Aplicadas, Santiago de Chile, Chile
*Corresponding author. e-mail address: asalles@instituteofneuroethics.org

Contents

1. Introduction	101
2. Anthropomorphism	103
3. Social chatbots	104
4. Two strategies	106
5. Pragmatic strategy: Some objections	108
6. Conclusion	114
Funding	114
References	114

Abstract

In this chapter, we are concerned with anthropomorphism in social AIs, particularly social chatbots. While not embodied in terms of having a human-like appearance, these chatbots are designed to induce attribution of intentionality and agency to engage users in enhanced interactions. Here we identify and present two strategies used to address the legitimacy of the anthropomorphisation, the intentional creation of human-like traits, of AIs: ontological and pragmatic. We further review some objections to pragmatic attempts to justify the anthropomorphisation of AIs. It is not our goal to argue in favour or against anthropomorphising social chatbots. Rather we examine some persistent concerns and call for more attention to and further research and reflection on the ethical, psychological, and ontological assumptions underlying them.

1. Introduction

Anthropomorphism, in general terms, refers to the attribution of human-like traits to non-human entities such as animals, natural and supernatural phenomena, and technology (Epley, Waytz, & Cacioppo, 2007). Such attribution often includes feelings, mental states, intentionality, and behaviours (Marchesi et al., 2019; Wayts, Cacioppo, & Epley, 2014).

Anthropomorphism has received a great deal of attention in the context of AI. Several authors have critically examined this phenomenon, its causes, underlying mechanisms, and potential and actual social and ethical impacts (Nyholm, 2020; Nyholm, 2023; Proudfoot, 2011). Multiple articles explore the pervasiveness of anthropomorphic thinking across different domains, including official documents and ethics guidelines to the research community and the general public (Deroy, 2023; Salles, Evers, & Farisco, 2020). Some authors have investigated the experiential and cultural differences that underlie anthropomorphism (Spatola & Wykowska, 2021, Spatola, Marchesi, & Wykowska, 2022). Others have explored the role of external factors in inducing anthropomorphism, from the use of anthropomorphic framing and narratives that shape public interpretation of AI (Darling, 2017; Ryan, 2020; Watson, 2019) to anthropomorphic design in social AI devices for a variety of purposes (Zlotowski, Proudfoot, Yogeeswaran, & Bartneck, 2015), and have shed light on potential consequences, including the attribution of moral standing and responsibility to anthropomorphised AIs. With the development, deployment, and increasing use of large language models designed to generate human-like responses to diverse outputs, anthropomorphisation and anthropomorphic attribution remain topics of attention, particularly considering the extent to which they may shape the interaction of people with those models and their subsequent behaviour.

Here, we are concerned with anthropomorphism in social AIs, particularly social chatbots (SCs). While not embodied in terms of having a human-like appearance, these chatbots are designed to induce attribution of intentionality and agency to engage users in enhanced interactions. Of course, many of the issues raised by anthropomorphism in these SCs are like those discussed in relation to anthropomorphised robots, so much of the existing research in social robotics is quite useful to identify the issues. The increasing availability of SCs, however, gives a sense of urgency to the need to address the concerns raised. Here, we identify and outline two strategies used to address the legitimacy of the anthropomorphisation, the intentional creation of human-like traits, of AIs: ontological and pragmatic. We review some objections presented to pragmatic attempts to justify the anthropomorphisation of AIs. It is not our goal to argue in favour or against anthropomorphising SCs. Rather we present what appear to be persistent concerns and call for more attention to and further research and reflection on the ethical, psychological, and ontological assumptions underlying them.

2. Anthropomorphism

The human tendency to anthropomorphise is a psychological phenomenon supported by a set of socio-cognitive mechanisms that are both reflective and non-reflective and are acquired during childhood (Perez-Osorio & Wykowska, 2020, Urquiza-Haas & Kotrschal, 2015). In a seminal article, Epley and colleagues conceptualise anthropomorphism as an inductive inference that relies on three co-determinants: a bottom up cognitive determinant comprising readily accessible general knowledge about the self in particular and humans in general, and two top down motivational factors: the need to interact effectively with non-human agents, which the authors call "effectance" motivation, and the need to establish social connections with others, dubbed "sociality" motivation by the authors (Epley et al., 2007; Epley, 2018). From this perspective, anthropomorphism becomes a fast, explanatory, and predictive strategy to understand and connect with non-human entities reducing the complexity and uncertainty of the interaction with them. As an automatic reaction, it is a less cognitively costly approach to interacting with the environment.

The three internal co-determinants account for the variety of anthropomorphism across humans but are far from uniform. Indeed, beyond them there are important individual differences which are determined by external dispositional, situational, cultural, and developmental variables (Bartz, Tchalova, & Fenerci, 2016; Epley et al., 2007), that is, additional triggers. For example, framing and human-like morphological or behavioural features in AIs (Fink, 2012) are likely to activate anthropomorphic reactions, and anthropomorphism tends to be more prevalent in some sociocultural backgrounds than others (Gal, 2020; Spatola et al., 2022). Importantly, as people engage in more demanding cognitive processing, they are likely to be less dependent on readily available anthropomorphic information and more receptive to alternative sources of information (Epley et al., 2007; Spatola & Wykowska, 2021).

It should be noted that some authors warn against assigning too much weight to anthropomorphic beliefs in the attribution of anthropomorphic features even if those features may sometimes lead to the formation of such beliefs (Damiano & Dumouchel, 2018). Airenti (2018) sees anthropomorphism as a phenomenon to be understood as a type of communicative interaction, "a way of relating with a non-human entity by addressing it as if it were a partner in a communicative situation" (Airenti, 2018, p. 8).

3. Social chatbots

Within AI, anthropomorphism has typically been addressed in relation to social robots which can not only simulate human speech but also mimic human appearance and behaviour. At present, it is gaining attention in relation to a key category of generally disembodied AIs: chatbots, that is, software applications that process human language and respond automatically by using text, voice, graphics, gestures, and other modes of communication that enable them to have social conversations with users.

Different kinds of chatbots can be distinguished by their functions and capabilities (Skjuve, 2021). Non-task-oriented models are designed for more open-ended conversations and are intended to either entertain or provide companionship, whereas task-oriented systems aim to achieve specific goals such as providing information or access to services (Mazur, Rzepka, & Araki, 2012). Virtual assistants or digital assistants, such as *Siri* and *Alexa*, are sophisticated, interactive, and personalised task-oriented chatbots. They can provide recommendations and anticipate needs by applying predictive intelligence and analytics based on user profiles and past user behaviour. Moreover, a distinction can be made between rule-based and machine-learning-based chatbots (Viduani, Cosenza, Araújo, & Kieling, 2023). Rule-based chatbots are programmed with a set of more or less complex rules which determine answers to given inputs. These are most efficient in narrow conversation domains, which is why task-oriented systems such as therapeutic chatbots are rule-based. In turn, machine learning-based chatbots learn from the analysis of a large body of desired interactions. The increasing availability of large amounts of natural language data on the internet and from digital media facilitate this type of training, allowing for improvement overtime and resulting in chatbots that are able to engage in fairly expressive conversations on a wide variety of domains.

In some conversational AIs, mimicking human traits and interactions is circumstantial or not essential to their functions. This is true for most task-oriented models. Although recent research shows that the manifestation of human-like traits may improve users' experience, those AIs could still achieve their function even if not mimicking humans in their interaction with users. In contrast, there is a sub-category of conversational AIs referred in the literature as SCs for which anthropomorphic traits are considered fundamental aspects of their function (Milne-Ives, De Cock, & Lim, 2020; Van Mezel, Cores, & Antheunis, 2021). Even further, some chatbots such as Replika, for example, exhibit identity traits, such as a specific personality, gender, visual appearance, and a unique set of emotional manifestations, which further facilitates their anthropomorphising.

SCs include task-oriented models such as artificial therapists (e.g. Wysa and Woebot Health) designed to address specific mental-health issues through human-like interaction with patients, and non-task-oriented models intended to embody a personal bond of the user, such as a friend, a sibling, or a romantic partner (e.g. Replika, Anima, Kuki and Xiaoice). These SCs are therefore designed to trigger the attribution of human-like traits, unlike other chatbots such as ChatGPT that specifically discourage anthropomorphisation through texts warning about their lack of human traits.

AI in general and chatbots in particular are increasingly being adopted in a variety of contexts for different purposes and their benefits and challenges are widely discussed (Abd-Alrazaq, Bewick, & Farragher, 2019; Milne-Ives et al., 2020; Sedlakova & Trachsel, 2023). Within mental health, chatbot use ranges from training, diagnosis, symptom screening, and management of several disorders to delivery of care (Boucher et al., 2021). While autism, depression, and anxiety are the most frequently addressed conditions, applications for suicide risk, substance abuse, post-traumatic stress disorder, stress, dementia, and acrophobia have been developed as well (Boucher et al., 2021; Viduani et al., 2023). Even when acknowledging limitations and the need for further research to fully understand the effectiveness of these tools, several studies observe that SCs show promising results in a variety of contexts. These include supporting physical and cognitive activities, increasing positive affect and perceived stress, promoting healthy behaviours, reducing feelings of loneliness, and helping the development of social abilities (Abd-Alrazaq et al., 2019; Fitzpatrick, Darcy, & Vierhile, 2017; Skjuve, 2021; Ta et al., 2020). Moreover, in a context that is characterised by a shortage of mental health professionals and by significant disparities, the almost constant availability of chatbots could contribute to equity by offering support to socio-economically vulnerable patients who would otherwise have no or only limited access to mental healthcare. Chatbots could encourage interaction in the case of patients living in communities where mental illness is stigmatised, and may reduce the workload of already overwhelmed caregivers (Ali, Zhang, Tauni, & Shahzad, 2023; Kretzschmar et al., 2019; Vaidyam, Wisniewski, Halamka, Kashavan, & Torous, 2019). Despite these promises, there are many ethical concerns about the use of such chatbots, relating for instance to privacy and safety (Luxton, 2020). The fact that some of these often unsupervised therapeutic tools might interact with psychologically vulnerable users that are unable to self-regulate their use also raises concerns about potential negative impacts, for example, over-attachment or over-reliance

(Kretzschmar et al., 2019; Vaidyam et al., 2019) or emotional dependence, addiction, and new forms of stigma (Laestadius, Bishop, Gonzalez, Illenčík, & Campos-Castillo, 2022; Xie, Pentina, & Hancock, 2023). Our focus here, though, is with the issues raised by anthropomorphisation, that is, the engineering of conversational AIs with human-like traits that are expected to trigger anthropomorphic reactions in users. We focus next on two strategies used to discuss its legitimacy.

4. Two strategies

The general question of whether the anthropomorphising of AIs is problematic has been widely debated. An ontologically oriented frame addresses it by focusing on the extent to which AIs have or may soon possess traits that are relevantly similar to human traits. If they do, then anthropomorphising them is not wrong.

However, despite continued collaboration between neuroscience and AI, significant advances in AI research, and structural affinities in brain inspired AI, the case for the ontological similarity between AI devices and humans is presently rather weak. AI research has consistently drawn inspiration from biological brains, especially the human brain. Indeed, some consider natural intelligence a blueprint for AI (Hassabis, Kumaran, Summerfield, & Botvinick, 2017; Summerfield, 2023). However, the cognitive capacities and skills of AIs are different from those in biological systems (Farisco, Evers, & Salles, 2020) and even when AIs are able to achieve goals, express behavioural flexibility, and innovation capacity, their goals and needs are quite different from those in biological systems (Farisco et al., 2023) which typically result from emotional interaction with the environment.

It is essential to be mindful that many challenges identified in building human-like AIs are technical and likely to be overcome as research and innovation progress. Indeed, it is not logically impossible that AIs will not just display but even possess some of the traits that humans have and display, for example, consciousness (Chalmers, 2023; Pennartz, Farisco, & Evers, 2019). Still, even if in the future AIs possess traits like general intelligence and consciousness, this does not necessarily imply that these traits will be human-like (Nyholm, 2023; Salles et al., 2020). The issue of whether anthropomorphism would be appropriate on ontological grounds would therefore still remain.

Given the uncertainty regarding an ontological match, an alternative strategy proposes that rather than relying on ontological considerations to justify anthropomorphism in AI, the focus should be on practical utility, including its potential contribution to non-moral and moral goods. According to this approach, the justification of anthropomorphising depends on whether it achieves desired goals. In some cases, the main goal might be to better understand and work with AI systems, that is what Epley and colleagues call effectance motivation. A recent study purports to show that participants successfully predicted whether a classifier would label an image correctly by using their own perceptions of the images as first-approximation proxies, a strategy that the authors called effective anthropomorphism (Bos, Glasgow, Gersh, Harbison, & Paul, 2019). There are many contexts in which the main goal might be to effectively work with AI systems, for example in medicine, to improve both medical research and applications (Starke & Poppe, 2022). On the pragmatic strategy, if implicit or explicit anthropomorphisation makes this goal more reachable by enhancing understanding and prediction of how AI systems work, then such anthropomorphism is a legitimate strategy.

However, this kind of anthropomorphic thinking is different from the one that interests us here mostly driven by the need for social connection, that is what Epley and colleagues call sociality motivation. In cases where the goal is for AI systems to socially interact with human users to better assist and care for them, the pragmatic strategy holds that if anthropomorphisation facilitates social interaction, provides better support, positively impacts moods, and generally increases wellbeing and efficient communication, then building anthropomorphic traits in those AIs would be both justified and desirable. Indeed, one important goal of social robotics research has been to endow robots with the human cues necessary to enhance familiarity and social interactions (Coeckelbergh, 2022; Damiano & Dumouchel, 2018; Fink, 2012; Zlotowski et al., 2015).

This idea underlies the deliberate anthropomorphisation of chatbots. Recent efforts focus on developing affective computing techniques and generating empathetic and emotional responses in SCs to elicit empathy and emotional responses constitutive of human relationships (Skjuve, 2021). Anthropomorphic conversational styles are expected to allow social AIs to satisfy some fundamental human needs, from communication to affection and social belonging (Zhou, Gao, Li, & Shum, 2020) which are considered significant for psychological well-being, integrity, and mental health.

According to the pragmatic view, anthropomorphisation is justified insofar as it produces the desired effect. In consequence, much research has aimed to understand how people react and relate to conversational SCs

(Croes & Antheunis, 2021; Hu, Lu, & Gong, 2021; Skjuve, 2021) and how relationships between humans and AIs form. Research continues to be conducted on the extent to which anthropomorphisation of chatbots truly enhances users' experiences (Hu et al., 2021; Konya-Baumbach, Biller, & Von Janda, 2023) and on which traits and cues induce the greatest sense of comfort. Several studies observe that the relationship formed between robots/chatbots and users may have a positive impact on users in various contexts (De Visser et al., 2016; Roy & Naidoo, 2021). However, cases where empirical analysis and examination show that anthropomorphic cues might trigger feelings of discomfort, lead users to disengage from reality (Sparrow & Sparrow, 2006), or cause harmful attachment, emotional dependence, or anxiety (Xie et al., 2023) would count against such anthropomorphising.

5. Pragmatic strategy: Some objections

Even in the best of cases, namely if anthropomorphisation of SCs achieves the goal of inducing a relationship that benefits users, the pragmatic strategy has faced criticism. For its objectors, the concern is not about utility or the potential effectiveness of anthropomorphism in improving users' well-being in a specific context. Instead, it revolves around the types of relationship formed between humans and artificial devices, relationships that they consider deceptive, qualitatively inferior, and a potential threat to humanness. Importantly, while the pragmatic strategy advocates for setting ontological considerations aside, a review of these concerns shows that they ultimately rest on ontological considerations. Explicit or implicit mentions of the ontological difference between artificial entities and humans underlie these concerns, especially with view to their capacity for emotions and interpersonal relationships. These different underlying assumptions often result in the two sides talking past each other.

The concern about alleged deception in human-artificial device interaction posits that chatbots displaying "empathy", offering positive reinforcement, having a "personality" with various apparent feelings, opinions, views, or expressing "concern" are deceiving users in morally significant ways (Montemayor, Halpern, & Farweather, 2022; Sedlakova & Trachsel, 2023).

From a pragmatic standpoint, the morally problematic nature of this deception would depend on the extent to which it promotes the desired goal, that is users' well-being (Wangmo, Lipps, Kressig, & Ienca, 2019).

However, this does not satisfy critics of the pragmatic strategy. We can, then, consider two additional ways to address the concern about deception: One questions whether a belief in the human-like nature of SCs is always involved when people interact with them. The other explores whether such belief always constitutes deception and, if so, whether it is morally problematic.

In a recent paper, Eva Weber-Guskar takes the first approach. She challenges the idea that users of social AIs necessarily believe that AIs have human-like traits such as emotions and are thus being deceived. Weber-Guskar notes that people might engage in a relationship, even an affective one, knowing that such relation is with an object and not a human. She offers two possibilities for avoiding the description of these types of relationships as deceptive. One is to understand users' interaction with social AIs as shaped by "imaginative perception", similar to emotional interaction with fiction. The second, compatible with much research on anthropomorphism discussed in section II, is to understand the attribution of human-like characteristics as a way to make sense of, interact, and get along with the system by developing and adopting what Daniel Dennett termed the "intentional stance" (Perez-Osorio & Wykowska, 2020; Weber-Guskar, 2021).

There is empirical support for the idea that not all users of SCs are actually deceived. Several authors note that some users choose to enter into affective relationships with companionship chatbots while specifically recognising their artificial nature (Brandtzaeg, Skjuve, & Folstad, 2022; Skjuve, 2021; Ta et al., 2020). This is also reported by a few users themselves (Jensen, 2023; Verma, 2023). Even further, it has been noted that the lack of similarity between users and chatbots, that is the very artificiality of chatbots, might actually strengthen users' feelings of trust and comfort (Ta et al., 2020), suggesting that some users might not only not be deceived, but also find the ontological difference particularly reassuring.

However, the observation that *sometimes* and in some cases there is no obvious deception, does not imply that users can always easily identify and understand that they are interacting with algorithms. So, the question remains: is giving cues of states that AIs do not necessarily possess morally wrong?

In a recent paper addressing AIs' deception, John Danaher notes that the tendency in the literature to conflate different forms of possible robotic deception complicates the discussion (Danaher, 2020). He introduces a distinction relevant to anthropomorphisation, namely between superficial state deception and hidden state deception. Superficial state deception occurs when a robot suggests that it has features that it does not have, while hidden state deception involves hiding abilities and capacities that it actually possesses. Determining what type of deception is involved is helpful in determining its moral quality. On ethical behaviourist grounds Danaher argues that we are epistemically justified in attributing mental states and capacities to AIs when the relevant set of behavioural cues is rich enough and consistent. This is why he holds that such instances of superficial state deception are "not an ethically worrisome form of deception in most cases" (p. 117) while hidden state deception can constitute a form of betrayal.

While Danaher specifically tries to avoid the ontological debate, the idea that the actual inner states of AIs do not matter to identify superficial state deception has been challenged (Musial, 2023) and concerns about deception in social AIs in general continue to be significantly shaped by ontological considerations about social AIs and their capacities (Saetra, 2020). While the lack of an ontological match is true with view to current models, there are many unknowns about models in the future. So, can we do what the pragmatic strategy proposes and set ontological considerations aside, focusing for instance on the *value or meaning* of human–chatbot relationships to compensate for at least some types of chatbot deception?

Critics of the pragmatic attempt to justify anthropomorphisation do not believe in such compensation. They argue that even if, for the sake of argument, we agree that, in many cases, SCs users are not deceived, and even if we grant that in cases of deception, such deception is not necessarily morally problematic, the resulting human–artificial entity relationship itself is deficient and less meaningful.[1] This deficiency is attributed to the absence of emotional reciprocity. In the context of mental health and healthcare the argument is that while mental health chatbots may be able to possess some form of cognitive empathy, detecting or recognising emotional mental states, they are in principle incapable of proper emotional or motivational empathy, requiring emotions that lead to empathic concern for others and deemed necessary to satisfy patients' needs (Montemayor et al., 2022).

Indeed, discussions over the ethics of chatbots in mental health and healthcare in general acknowledge the differences in how users interact and form relationships with AIs compared to interactions with other humans.

---

[1] One of us (AS) together with colleagues has argued that there might be some contexts (for example, elderly care) in which for actions to be morally sound and beneficial they might have to be the result of some traits that, if absent in AIs, would make their use morally problematic (Farisco, Evers, & Salles, 2020). This view raises a number of additional issues, however, including how to assess the moral quality of actions, that we do not address here.

Despite the recognition that chatbots cannot be regarded as genuine agents or be involved in genuine human relationships, proponents argue that they can nonetheless contribute to the user's self-understanding and self-acceptance (Grodniewicz & Hohol, 2023). For accomplishing this, the chatbot itself does not need to have self-understanding, and neither does a therapist. Therapists often play only a facilitating role, where understanding and integration of information and emotions are performed by the patient, who is the active agent of change (Holohan, Buyx, & Fiske, 2023). Some commentators even argue that therapeutically efficacious conversations do not require a real agent but rather the patient's phenomenological experience of being emotionally supported, understood, and encouraged to engage in introspection (Hurley, Lang, & Smith, 2023). If so, the lack of emotional reciprocity might not have the moral significance that detractors of the pragmatic strategy believe it has.

However, in the case of friendship companion chatbots, critics argue that the significance of mutually caring and affective relationships, which depend on emotional capacities absent in AIs, is especially relevant. Unlike mental health chatbots that primarily offer strategies to monitor and manage health issues, companion chatbots are intended to elicit emotional responses, facilitate emotional bonds with users, and provide empathetic support for humans in stressful and everyday contexts. Consider Replika again, advertised as a friend that can coach users for managing difficult emotions. Through a machine learning approach, it develops a tailor-made personality for its users by learning from its interaction with them, and creates an apparent bi-directional relationship where each party has their own "subjectivity" (Skjuve, 2021; Ta et al., 2020; Xie et al., 2023). Critics argue that these relationships do not lead to genuine bonds with users, and the potential meaning or value of the resulting human–chatbot relationship cannot justify anthropomorphisation, as meaningful relationships have ontological conditions: in particular, that the parties possess actual emotions with which to reciprocate.

Notably, a few empirical studies report that while some users described their friendship with Replika as long term, they did not perceive it as "real", and that even when they use terms as "reciprocity" when describing their relationship with the device, such reciprocity was described in terms of sharing ideas and helping each other (Brandtzaeg et al., 2022) rather than in terms of shared experiences and emotional interactions. The question, then, becomes: is the objector correct in holding that those relationships are illusory and meaningless?

One way to address this concern is by questioning the underlying understanding of affective relationships. Weber-Guskar (2021), for example, argues that mutuality and emotional reciprocity, while desirable, are not necessary for good affective relationships. She points out instances of positive personal affective relationships that are one sided and instances of affective relationships with partners who exhibit some desirable behaviour even when they might not have feelings at all. Since those human-to-human relationships, she holds, are neither illusory nor meaningless, there is no reason to think that they would be illusory and meaningless in the case of human–chat relationship. Importantly, Weber-Guskar emphasises the need to disentangle ontological and moral considerations, noting that arguing for the possibility of meaningful one-sided affective relationships is not to claim that they are in every respect as good as relationships with emotional reciprocity.

There is an alternative way to address concerns about lack of mutuality or emotional involvement, however: rather than trying to show the potential similarity between human–chatbot relationship and some human–human relationships, some propose that we go beyond anthropocentric understandings of good relationships and consider instead that in some cases interacting with technology might entail a new kind of intimate connection, a new form of personalised relationship that can be meaningful in an altogether different way (Brandtzaeg et al., 2022). Going beyond the distinction between real or authentic and fake relationships, the focus should be put on the extent to which those relationships offer the possibility of deeper knowledge and potentially even moral growth (Damiano & Dumouchel, 2018). For instance, some users report that their relationship with companion chatbots contributed not just to their self-understanding but also to their self-acceptance. The non-judgmental environment allows them to own traits that they have hidden out of fear of social or self-stigmatisation (Jensen, 2023; Verma, 2023). If so, the desirability of these relationships rests on the opportunity they provide users to explore and deepen their own emotions, offering the possibility to enhance the process of developing users' personality, values, and self-knowledge even if they are not genuinely human interactions.

The above, however, raises additional concerns voiced by critics of the pragmatic strategy that look at these relationships in the context of their potential impact on more specific groups of people and, ultimately, humanness at large. Indeed, some commentators view the readiness to engage in these different albeit personalised relationships as a manifestation of what Turkle has called the "robotic moment" (Turkle, 2011). Research indicates that human beings tend to interact with embodied or, in the case of chatbots, disembodied AIs as if they had some agency, even if they do not necessarily see AIs as subject of experiences (Geiselman, Tsourgianni, Deroy, & Harris, 2023). This strongly suggests human complicity in this illusion of a meaningful connection. Critics find this complicity problematic for two reasons. First, it may lead to a misplacement of emotional resources. Second, it manifests a problematic position in which humans appear willing to consider robots as potential friends, confidants, and even romantic partners, even when aware that artificial entities cannot reciprocate. From this perspective, the acceptance of these relationships might have significant implications both individually and on humanness at large.

Individually, accepting the legitimacy of these relationships may distort human priorities and potentially facilitate the inappropriate replacement of humans, for example, in loving relationships, negatively affecting specific others towards whom humans might have moral duties (Weber-Guskar, 2021). Moreover, there is the concern that accepting and engaging in these kinds of relationships might result in more vulnerable and diminished human beings. For critics, instead of offering a possibility for human empowerment and self-knowledge, the use of artificial devices expressing emotions they don't genuinely possess, and the deliberate engagement with them, might have serious epistemic and metaphysical implications. Epistemic implications could arise as humans get used to technocratic ways of thinking about humanness (Porra, Lacity, & Parks, 2020), potentially obscuring people's understanding of the distinction between things and humans. Metaphysical implications are suggested in the potential diminishment of humanness itself (Turkle, 2011).

While these concerns are worthy of attention, more empirical and theoretical research as well as engagement with diverse publics is needed to determine the moral strength of these types of concern. More empirical research is needed because, even if possible, it is not self-evident that accepting the meaningfulness of some human–chatbot relationships will distort people's priorities and lead them to prefer artificial devices over humans. In fact, some suggest that as we continue to value human traits such as for example, empathy, the availability of human–chatbots interactions might do the opposite, and "prompt a shift towards more direct, genuine, in person interactions" (Perry, 2023). In any case, it seems that continuing to study and assess the psychological impacts of SCs in diverse settings and different cultural contexts, and understanding risk factors, is key in determining the way forward.

On a theoretical level, more research is needed as well. While this might be an all too obvious recommendation, it is an important one, considering the increasing availability of these innovations and the oftentimes hasty description of the ethical issues they raise. Such research would benefit from interdisciplinarity and should include the identification of the different types of claims, including ethical, psychological, and ontological ones, as well as a deeper reflection on some basic concepts whose meaning is often taken for granted. Concerns about our beliefs about humans and humanness being threatened by social interaction with these devices should prompt a careful examination of what humanness consists in, how it is perceived, what kind of alteration leads to its diminishment, and why (Salles, 2021). Further reflection is also needed on how to conceptualise human diminishment in ontological and moral terms, as well as an exploration of whether all human-chatbot interactions, regardless of their quality, lead to morally undesirable human diminishment.

6. Conclusion

In this short contribution we have presented two strategies that play an important role in the discussion over anthropomorphised AIs. Our aim has not been to argue for or against the legitimacy of anthropomorphisation of SCs but rather to highlight some of the relevant considerations that continue to shape the conversation on the desirability of social AIs. Given their increasing use and potential impact, it is critical to recognise that there is still work to be done to unveil and better understand the benefits of these devices and the challenges they present. Fully addressing these issues promises to be an ongoing task requiring continued empirical exploration, research, and reflection not just on impacts but also on underlying assumptions and ethical and ontological commitments.

Funding

AW-P research was funded by Fondecyt Iniciacion Project N 11220327 (Chilean National Agency for Research and Development).

References

Abd-Alrazaq, A. A., Bewick, B. M., Farragher, T., et al. (2019). Factors that affect the use of electronic personal health records among patients: A systematic review. *International Journal of Medical Informatics, 126*, 164–175.

Abd-Alrazaq, A. A., Alajlani, M., Alalwan, A. A., Bewick, B. M., Gardner, P., & Househ, M. (2019). An overview of the features of chatbots in mental health: A scoping review. *International Journal of Medical Informatics, 132*, 103978.

Airenti, G. (2018). The development of anthrpomorphism in interaction: Intersubjectivity, imagiantion, and theory of mind. *Frontiers in Psychology, 9*, 2136.

Ali, F., Zhang, Q., Tauni, M. Z., & Shahzad, K. (2023). Social chatbot: My friend in my distress. *International Journal of Human-Computer Interaction.*

Bartz, J., Tchalova, K., & Fenerci, C. (2016). Reminders of social connection can attenuate anthropomorphism: A replication and extension of Epley, Akalis, Waytz and Cacioppo (2008). *Psychological Science, 27*(12), 1644–1650.

Brandtzaeg, P. B., Skjuve, M., & Folstad, A. (2022). My AI friend: How users of a social chatbot understand their human-AI friendship. *Human Communication Research, 48,* 404–429.

Bos, N., Glasgow, K., Gersh, J., Harbison, I., & Paul, C. L. (2019). Mental models of AI-based systems: User predictios and explanation of image classification results. In *Proceedings of the human factors and ergonomics society annual meeting* (pp. 184–188). Los Angeles, CA: SAGE.

Boucher, E. M., Harake, N. R., Ward, H. E., Stoeckl, J. V., Minkel, J., Parks, A., & Zilca, R. (2021). Artificially intelligent chatbots in digital mental health interventions: A review. *Experto Review of Medical Devices, 18*(sup1), 37–49.

Chalmers, D. (2023). *Could a large language model be conscious?* Retrieved from PhilPapers. org: ⟨https://philpapers.org/archive/CHACAL-3.pdf⟩.

Coeckelbergh, M. (2022). Three responses to anthrpomnorphism in social robotics: Towards a critical, relational, and hermeneutical approach. *International Journal of Social Robotics, 14,* 2049–2061.

Croes, E., & Antheunis, M. L. (2021). Can we be friends with Mitsuku? A longitudinal study on the process of relationship formation between humans and a social chatbot. *Journal of Social and Personal Relationships, 279–300.*

Damiano, L., & Dumouchel, P. (2018). Anthropomorphism in human–robot co-evolution. *Frontiers in Psychology, 9.* https://doi.org/10.3389/fpsyg.2018.00468.

Danaher, J. (2020). Robot betrayal: A guide to the ethics of robotic deception. *Ethics and Informationa Technology, 117–128.*

Darling, K. (2017). "Who is Johnny?" Anthropomorphic framing in human-robot interaction, integration, and policy. In A. K. P. Lin (Ed.). *Robot ethics.* New York: Oxford University Press.

Deroy, O. (2023). The ethics of terminology: Can we use human terms to describe AI? *Topoi, 42,* 881–889.

De Visser, E., Monfort, S., Mc Kendrick, R., Smith, M., Mc Knight, P., Krueger, F., & Parasuraman, R. (2016). Almost human: Anthropomorphism increases trust reliance in cognitive agents. *Journal of Experimental Psychology, 331–348.*

Epley, N., Waytz, A., & Cacioppo, J. (2007). On seeing human: A three-factor theory of anthropomorphism. *Psychological Review, 114*(4), 864–885.

Epley, N. (2018). A mind like mine: The exceptionally ordinary underpinnings of anthropomorphism. *JACR, 3*(4), 591–598.

Farisco, M., Baldasarre, G., Cartoni, E., Leach, A., Petrovici, M. A., Rosemann, A., & Van Albada, S. J. (2023). A method for the ethical analysis of brain-inspired AI. *Under review.*

Farisco, M., Evers, K., & Salles, A. (2020). Towards establishing criteria for the ethical analysis of AI. *Science and Engineering Ethics.*

Fink, J. (2012). Anthropomorphism and human likeness in the design of robots and human-robot interaction. *Social Robotics. ICSR 2012. Lecture notes in computer science* (pp. 199–208). Berlin: Springer.

Fitzpatrick, K. K., Darcy, A., & Vierhile, M. (2017). Delivering cognitive behavior therapy to young adults with symptoms of depression and anxiety using a fully automated conversational agent (Woebot): A randomized controlled trial. *JMIR Mental Health, 4.*

Gal, D. (2020). Perspectives and approaches in AI ethics: East Asia. In M. P. Dubber (Ed.). *The Oxford handbook of ethics of AI* (pp. 607–624)New York, NY: Oxford University Press.

Geiselman, R., Tsourgianni, A., Deroy, O., & Harris, L. (2023). Interacting with agents withiout a mind: The case for artificial agents. *Current Opinion in Behavioral Sciences, 51,* 101282.

Grodniewicz, J. P., & Hohol, M. (2023). Therapeutic conversational artificial intelligence and the acquisition of self-understanding. *AJOB Neuroscience,* 59–61.

Hassabis, D., Kumaran, D., Summerfield, C., & Botvinick, M. (2017). Neuroscience-inspired artificial intelligence. *Neuron, 95*(2), 245–258.

Holohan, M., Buyx, A., & Fiske, A. (2023). Staying curious with conversational AI in psychotherapy. *AJOB Neuroscience, 23*(5), 14–16.

Hu, P., Lu, Y., & Gong, Y. (2021). Dual humanness and trust in conversational AI: A person centered approach. *Computers in Human Behavior.*

Hurley, M. E., Lang, B. H., & Smith, J. N. (2023). Therapeutic artificial intelligence: Does agential status matter? *AJOB Neuroscience, 23*(5), 33–35.

Jensen, T. (2023, March 9). An AI 'Sexbot' fed my hidden desires—And then refused to play. *Wired.*

Konya-Baumbach, E., Biller, M., & Von Janda, S. (2023). Someone out there? A study on the social presence of anthropomorphized chatbots. *Computers in Human Behavior.*

Kretzschmar, K., Tyroll, H., Pavarini, G., Manzini, A., Singh, I., & NeurOx Young People's Advisory Group (2019). Can your phone be your therapist? Young people's ethical perspectives on the use of fully automated conversational agents (chatbots) in mental health support. *Biomedical Informatics Insights.*

Laestadius, L., Bishop, A., Gonzalez, M., Illenčík, D., & Campos-Castillo, C. (2022). Too human and not human enough: A grounded theory analysis of mental health harms from emotional dependence on the social chatbot Replika. *New Media & Society.*

Luxton, D. (2020). Ethical implications of conversational agents in global public health. *Bulletin of the World Health Organization, 98,* 285–287.

Marchesi, S., Ghiglino, D., Ciardo, F., Perez Osorio, J., Baykara, E., & Wykowska, A. (2019). Do we adopt the intentiohal stance toward humanoid robots? *Frontiers in Psychology, 10,* 450.

Mazur, M., Rzepka, R., & Araki, K. (2012). Chatterbots with occupation-between non task and task oriented conversational agents. In *Linguistic and cognitive approaches to dialogue agents AISB/IACAP symposium* (pp. 61–66).

Milne-Ives, M., De Cock, C., Lim, E., et al. (2020). The effectiveness of artificial intelli-gence conversational agents in health care: Systematic review. *Journal of Medical Internet Research, 22*(10), e20346. https://doi.org/10.2196/20346.

Montemayor, C., Halpern, J., & Farweather, A. (2022). In principle obstacles for empathic AI: Why we can't replace human empathy in healthcare. *AI & Society, 37,* 1353–1359.

Musial, M. (2023). Criticizing Danaher's approach to superficial state deception. *Science and Engineering Ethics, 29,* 31.

Nyholm, S. (2020). *Humans and robots: Ethics, agency and athropomorphism.* Lanham, MD: Rowman and Littefield.

Nyholm, S. (2023). Robotic animism: The ethics of attributing minds and personality to robots with artificial intelligence. In S. T (Ed.), *Animism and philosophy of religion* (pp. 313–340). Cham: Springer.

Pennartz, C., Farisco, M., & Evers, K. (2019). Indicators and criteria fo consciousness in animals and intelligent machines: An inside out approach. *Frontiers in Systems Neuroscience.*

Perez-Osorio, J., & Wykowska, A. (2020). Adopting the intentional stance toward natural and artificial agents. *Philosophical Psychology, 33,* 369–395.

Perry, A. (2023). AI will never convey the essence of human empathy. *Nature Human Behabiour.*

Porra, J., Lacity, M., & Parks, M. S. (2020). Can Computer Based Humsn-Likeness Endanger Humanness?—A philosophical and ethical perspective on digital assistants expressing feelings they can't have. *Information Systems Frontiers, 22,* 533–547.

Proudfoot, D. (2011). Anthropomorphism and AI: Turing's much misunderstood imitation game. *Artificial Intelligence, 175*(5-6), 950–957.

Roy, R., & Naidoo, V. (2021). Enhancing chatbot effectiveness: The role of anthropomorphic conversational styles and time orientation. *Journal of Business Research, 23–34.*

Ryan, M. (2020). In AI we trust: Ethics, artificial intelligence, and reliability. *Science and Engineering Ethics,* 2749–2767.

Saetra, H. S. (2020). The parasitic nature of social AI: Sharing minds with the mindless. *Integrative Psychological and Behavioral Science, 54,* 308–326.

Salles, A., Evers, K., & Farisco, M. (2020). Anthropomorphism in AI. *AJOB Neuroscience, 11*(2), 88–95.

Salles, A. (2021). Humanness: Some neuroethical reflections. In M. Hevia (Ed.). *Regulating neuroscience: Transnational legal challenges developments in neuroethics and bioethics* (pp. 1–17). Cham: Elsevier.

Sedlakova, J., & Trachsel, M. (2023). Conversational artificial intelligence in psychotherapy: A new therapeutic tool or agent? *AJOB Neuroscience,* 4–13.

Skjuve, M. F. (2021). My Chatbot companion—A study of human-chatbot relationships. *International Journal of Human-Computer Studies, 149.*

Sparrow, R., & Sparrow, L. (2006). In the hands of machines? The future of aged care. *Minds and Machines,* 141–161.

Spatola, N., & Wykowska, A. (2021). The personality of anthropomorphism: How th eneed for cognition and the need for closure define attitudes and anthrpomorphic attributions towards robots. *Computers in Human Behavior, 122,* 106841.

Spatola, N., Marchesi, S., & Wykowska, A. (2022). Different models of anthropomorphism across cultures and ontological limits in current frameworks. *Frontiers in Robotics and AI, 9,* 863319.

Starke, G., & Poppe, C. (2022). Karl Jaspers and artificial neural nets: On the relation of explaining and understanding artificial intelligence in medicine. *Ethics and Information Technology, 24,* 26.

Summerfield, C. (2023). *Natural general intelligence: How understanding the brain can help us build AI.* New York, NY: Oxford University Press,.

Ta, V., Griffith, C., Boatfield, C., Wang, X., Civitello, M., Bader, H., ... Loggarakis, A. (2020). User experiences of social support from companion chatbots in everyday contexts: Thematic analysis. *Journal of Medical Internet Resrach.*

Turkle, S. (2011). *Alone together: Why we expect more from technology and less from each other.* New York: Basic Books,.

Urquiza-Haas, E., & Kotrschal, K. (2015). The mind behind anthropomorphic thinking: Attribution of mental states to other species. *Animal Behaviour, 109,* 167–176.

Vaidyam, A. N., Wisniewski, H., Halamka, J. D., Kashavan, M. S., & Torous, J. B. (2019). Chatbots and conversational agents in mental health: A review of the psychiatric landscape. *The Canadian Journal of Psychiatry,* 456–464.

Van Mezel, M., Cores, E., & Antheunis, M. (2021). "I'm Here for You": Can social chatbots truly support their users? A literature review. In A. E. Flostadt (Ed.). *Chatobot research and design.* Cham: Springer.

Verma, P. (2023, March 30). They fell in love with AI bots. A software update broke their hearts. *Washington Post.*

Viduani, A., Cosenza, V., Araújo, R. M., & Kieling, C. (2023). Chatbots in the field of mental health: Challenges and opportunities. In R.-D.-P. F. I. Cavalcante Passos (Ed.). *Digital mental health: A practitioner's guide* (pp. 133–148). Cham: Elsevier.

Wangmo, T., Lipps, M., Kressig, R. W., & Ienca, M. (2019). Ethical concerns with the use of intelligent assistive technology: Findings from a qualitative study with professional stakeholders. *BMC Medical Ethics.*

Watson, D. (2019). The rhetoric and reality of anthropomorphism in artificial intelligence. *Minds and Machines, 29,* 417–440.

Wayts, A., Cacioppo, J., & Epley, N. (2014). Who sees human? The stability and importance of individual differences in anthropomorphism. *Perspectives on Psychological Science, 5*(3), 219–232.

Weber-Guskar, E. (2021). How to feel about emotionalized artificial intelligence? When robot pets, holograms, and chatbots become affective partners. *Ethics and Information Technology, 23*, 601–609.

Xie, T., Pentina, I., & Hancock, T. (2023). Friend, mentor, lover: Does chatbot engagement lead to psychological dependence? *Journal of Service Management,* 806–828.

Zhou, L., Gao, J., Li, D., & Shum, H. Y. (2020). The design and implementation of xiaoice, an empathetic social chatbot. *Computational Linguistics,* 53–93.

Zlotowski, J., Proudfoot, D., Yogeeswaran, K., & Bartneck, C. (2015). Anthropomorphism: Opportunities and challenges in human-robot interaction. *International Journal of Social Robotics, 7*, 347–360.

CHAPTER SEVEN

(Mis)decoding affect in the face and in the brain

Marco Viola*
Department of Philosophy, Communication, and Performing Arts, University of Roma Tre, Rome, Italy
*Corresponding author. e-mail address: marco.viola@uniroma3.it

Contents

1. QWERTY keyboards and affective science		119
1.1 Emotional faces as QWERTY keys		120
1.2 The QWERTY-like development of the psychoevolutionary theories of emotion		121
2. Decoding emotion from faces		125
3. Decoding emotion from (neurovascular proxies for) brain data		129
4. Summary and open problems		134
Acknowledgments		135
References		135

Abstract

Despite technical progress, automatic systems aimed at "decoding" a subject's affective states based on objective measures, such as patterns of facial movements or neural activity, are undermined by intricate epistemological and theoretical issues. Most of these systems rely on some principles from Paul Ekman's research on emotion and his taxonomy of the "Canonical Six" emotion categories. However, there is a growing consensus in affective science that these principles and categories require updating or even rejection. In this chapter, I illustrate some of these issues and discuss the risks that they may lead to misdecoding affective states.

1. QWERTY keyboards and affective science

In most countries using Latin alphabets, the layout of computer keyboards follows a similar pattern: starting from the top-left corner, you will find the same set of typographical characters, namely "Q", "W", "E", "R", "T", "Y", and so on. This layout, known as QWERTY after the first six letters, is not intrinsic to electronic computers. Instead, this arrangement of keys was originally developed to improve the efficiency of the earliest writing machines during the late 19th century. Modern computers have

Developments in Neuroethics and Bioethics, Volume 7
ISSN 2589-2959, https://doi.org/10.1016/bs.dnb.2024.02.002
Copyright © 2024 Elsevier Inc. All rights are reserved, including those for text and data mining, AI training, and similar technologies.

simply inherited this layout because their users were accustomed to that arrangement while typing—not because QWERTY keyboards are inherently efficient for typing on electronic computers. Although oversimplified, this narrative of technological (lack of) development is often invoked as a prototype to describe the notion of *path dependence*. In simpler terms, path dependence occurs when a certain feature of the world is primarily shaped by the historical context or the needs at some point in the past, rather than being optimally suited to present conditions.

This simplified story serves as a metaphor for two brief scenarios in which inertia plays a role in affective science: the evolution of emotion and the evolution of emotion theory.

1.1 Emotional faces as QWERTY keys

Around 1850, the renowned biologist Charles Darwin endeavored to demonstrate that natural selection, as opposed to intelligent design, is the primary driving force behind the form and behavior of organisms. According to his perspective, far from being separate from other animal species, humankind is merely one of the numerous potential outcomes of evolutionary history.

However, it was far from evident how this evolutionary account could explain the distinct expressiveness of the human face. After all, the intricate musculature of the human face offers an estimated 3.7×10^{16} possible combinations, allowing for a wide range of expressions (Lee, Anderson, Fernandez-Dols, & Russell, 2017).

The Scottish surgeon Charles Bell provided a straightforward and compelling explanation for the uniqueness of the human face: he posited that this intricate expressive apparatus had been intricately designed by a divine creator and bestowed upon humans to establish a universal means of communicating affective states. The simplicity of this argument presented a rather persuasive rationale for supporting creationism. In order to advocate for evolutionism, Darwin needed to provide a more robust narrative. This endeavor proved to be challenging and time-consuming, spanning several years. Ultimately, in 1872—15 years after the initial publication of *"The Origin of Species"*—he released *"The Expression of the Emotions in Man and Animals"* (Darwin, 1872). This book offers an evolutionary elucidation of what are commonly recognized as emotional expressions, grounded in three principles and substantiated by various sources of evidence. Particularly relevant to our discussion is the first principle, known as the *"Principle of Serviceable Associated Habits"*. This principle posits that emotional behaviors, such as baring our teeth when angry, are remnants of movements that once held immediate functional value for our distant ancestors. In this case, this behavior could be linked to preparing to bite in response to a threat.

Similar to the arrangement of QWERTY keyboards, the explanation of present-day organisms' affective behavior, including what we commonly term "communicative" behavior or emotional expressions, is to be found in the annals of the past. While certain adjustments remain feasible as the social signaling functions of certain movements gain prominence (see Shariff & Tracy, 2011), the essence of this notion endures. This concept lies at the heart of a research paradigm frequently referred to as psychoevolutionary theories of emotion expression. Paradoxically, the history of this paradigm itself seems to be influenced by the phenomenon of path dependence.

1.2 The QWERTY-like development of the psychoevolutionary theories of emotion

Despite its huge initial editorial success, in the decades following the publication of "*Expression*", the reception of Darwin's theory was tepid. A possible reason might have been the influence of a behaviorist milieu in psychology at the beginning of the 20th century, as behaviorists were more inclined to ascribe the genesis of mental phenomena to learning rather than to the biological innateness that permeates Darwin's theory. At least, this was one of the many explanations suggested by Paul Ekman in his preface to the third edition of Darwin's book (Ekman, 1998).

Whatever the reasons for this dismissal of Darwin's ideas on emotion, in the second half of the last century the tide changed drastically, and many Darwinian motifs became decisively mainstream in the field of affective science. Indeed, this change of tide was largely courtesy of the very Paul Ekman—whose edition of "*Expression*", all things considered, was aimed at presenting himself as Darwin's intellectual heir (Plamper, 2015).

According to the psychoevolutionary paradigm inspired by Darwin, emotions are conceived as sort of hardwired behavioral toolkits aimed at providing fast and frugal, highly stereotyped solutions to recurrent evolutionary problems related to organisms' survival and reproduction. These solutions have been shaped by millennia of evolution, which helps explaining why they might sometimes be ill-tuned with some organisms' present environment. While many theorists diverged on some specific details, the theoretical core of this paradigm was largely endorsed by many scholars during the last decades of the 20th century (e.g. Griffiths, 1997; Panksepp, 1998; Tooby & Cosmides, 1990), and ultimately coalesced into a sort of "handbook truth".

Inspired by Silvan Tomkins, and in parallel with Carroll Izard, Paul Ekman contributed to this paradigm by articulating a research protocol that easily became an exemplar case in Kuhnian terms (Kuhn, 1970), that is the blueprint for further experiment within that paradigm. Several subjects were shown a series of photographs of faces contracting some specific patterns of muscles and asked to associate these pictures with some of the following pre-given emotion categories: anger, disgust, fear, happiness, sadness, surprise. The results of these "emotion recognition" experiments showed that people from several countries ascribed similar labels to the same pictures well above the level of chance (Ekman, Sorenson, & Friesen, 1969). Interestingly, the face-emotion link remained consistent also in preliterate subjects, the Fore of the New Guinea (Ekman & Friesen, 1971), speaking in favor of a biologically innate origin of the facial behavior-emotion link. To better construe this link, Ekman and Friesen (1978) also developed the Facial Action Coding System (FACS), an exhaustive list of all possible facial movements: the contraction of any group of facial muscles is represented by a number (together with a degree of intensity starting from its second edition).

In the following decades, despite methodological criticisms being no absent (see notably Russell, 1994; Fridlund, 2017), several researchers followed Ekman and Friesen's (and Tomkins and Izard's) footsteps, largely replicating their results, although they also highlighted the existence of more cultural variability than initially thought (Elfenbein & Ambady, 2002).

What about now? *Prima facie*, psychology has moved beyond the Ekmanian paradigm according to which "one facial pattern = one emotion category". Indeed, an increasing number of articles is replete with parricide attempts toward Paul Ekman, that is, recognizing his contribution to the field while at the same time inviting us to move forward. A recurrent criticism is that the photographic stimuli employed in emotion recognition studies tend to have poor ecological validity, as they often depict posed rather than spontaneous expressions. Aviezer, Bentin, Dudarev, and Hassin (2011) documented that the same face could be interpreted as expressing different emotions depending on bodily movements and other contextual cues. Cowen and Keltner (2021) claim that we must search for markers of emotion categories beyond the face, and that by doing so we might find way more than the traditional six categories figuring in Ekman's original list. More radically, the notable Theory of Constructed Emotions by Lisa Feldman Barrett (2017) claims that searching for stable "fingerprints" of emotion categories is a non-starter, as the boundaries of emotion categories are dictated by our cultural practices rather than being pre-defined by our biology.[1]

While we cannot indulge in a careful examination of these and other criticisms toward Ekman's account,[2] my hunch is that, as soon as we move from what scientists *claim* to what scientists (and engineers) *do*, these parricide attempts have not succeeded—at least not yet. In fact, although *in theory* many psychologists today would likely feel the urge to distance themselves from the simplistic view that "facial movement = emotion", *in practice* Ekman's legacy has become a kind of QWERTY layout for affective science, inasmuch its core commitments are deeply embedded in the very ways in which the scientific community designs experiments and develop the tools to perform them. For instance, think of the databases of facial expressions, that is large and standardized collections of static pictures (or more rarely, short video clips) of human faces while performing some movement that can be interpreted as expressing some emotion. The very existence of many such databases certifies the popularity of the core claim of Ekman's account, as they assume that a discrete emotion is encoded into the human face and can thus be read out of it straightforwardly. Having analyzed the stimuli used in all the papers on face perception published in a representative sample of psychology and neuroscience journal from 2000 to 2020, Dawel, Miller, Horsburgh, and Ford (2021) note that about 70% of the studies on the facial recognition of emotion adopted either Ekman's original photographs set or three other popular databases, namely the Karolinska Directed Emotional Faces, the NimStim, or the Radboud Database. But even those three databases, despite some technical improvements with respect to Ekman's original photos (e.g. the pictures are in color rather than in gray), share many ekmanian assumptions (and potential flaws). For instance, they are made by posed expressions. And indeed, even if recent

---

[1] In the earliest formulation of her theory, Barrett held that emotion is constructed based on conceptual interpretation of a given pre-conceptual affective state (see for instance Barrett & Bliss-Moreau, 2009). Like Russell (2003), another advocate of constructionism, Barrett construed this affective state in terms of two bipolar axes, valence (pleasure-displeasure) and arousal (activation-deactivation). However, in the most recent formulations of her theory, she seems no longer committed to this structure of affective states.

[2] For extended criticisms, see Russell (1994), Crivelli and Fridlund (2018) and Barrett et al. (2019).

databases that tried to address this issue by sampling spontaneous facial responses to induced emotions (e.g. Miolla, Cardaioli, & Scarpazza, 2022), almost no databases propose facial expressions for emotions beyond the six categories originally studied by Ekman that is anger, disgust, fear, happiness, sadness, surprise (though see Benda & Scherf, 2020).

Indeed, despite its origin being largely due to historical contingencies (see Ellsworth, 2014), the list of the six "basic" or "canonical" emotions later got reified and became a poorly disputed ontological axiom for many psychologists. This ontological dogma did not remain confined to psychology. Instead, it spilled over into other disciplines interested in affect. For instance, a recent study shows that the Ekman's original categories (plus "anxiety" and "pleasure") are the most likely emotion labels to be found in neuroscientific investigations of affect (Giannakoupolou et al., in preparation). Similarly, while psychologists are increasingly prone to reject or at least to overcome Ekman's theoretical legacy, engineers in charge of designing automated facial emotion recognition (FER) systems conceive them in terms of a task of pattern recognition, in which an algorithm must assign a given image (a face performing some muscular movements) to one of Ekman's emotion categories based on its resemblance to the patterns of facial movements stereotypically associated with them (see for instance Cohn et al., 2015), often decomposing facial movements into the movements codified by Ekman's FACS, a formal classification of facial movements (Ekman & Friesen, 1978).

In the remainder of this essay, I shall sketch the workings of two popular systems for decoding emotions, namely functional Magnetic Resonance Imaging (fMRI) data and FER technology (for a detailed overview of many epistemological issues of several methods for automated decoding affective states and a discussion of their ethical significance, see Stark & Hoey, 2021). In so doing, I shall try to highlight some epistemological issues that complicate their interpretation, with a special emphasis on the impact of Ekman's legacy.

Please note that, while on the one hand I am sincerely convinced that affective science would benefit by overcoming Ekman's legacy, on the other hand I am skeptical about some radical criticisms such as those that reject *any* patterned link whatsoever between facial movements and emotions (e.g. Barrett, 2017; Crivelli & Fridlund, 2018). In a slogan, I would refrain from throwing out the proverbial baby together with the bathwater. Most likely, to pave the way for post-QWERTY keyboards, we still need to tap some keys on our current QWERTY keyboards.

2. Decoding emotion from faces

While we need machines to "read" mental states based on neural data, neurotypical humans possess a natural ability to discern various mental states from facial cues–or, at the very least, to form impressions. Emotions figure prominently among the mental states that allow this kind of face-based mindreading.

The human face stands as one of the most attention-grabbing stimuli in our social environment. From our earliest days of life, we exhibit a preference for face-like patterns, directing more attention to them than to similar stimuli (Buiatti et al., 2019). This inclination likely stems from the evolutionary advantage of prioritizing information-rich surfaces like the human face. According to prevailing theoretical frameworks, upon perceiving a face or a facial-like pattern (referred to as *facial pareidolia*), neurocognitive mechanisms are set in motion to *detect* the presence of a face ("there is a they"). Subsequently, the visual characteristics of the face are processed spontaneously by (at least) two relatively independent neurocognitive mechanisms: one for facial identity ("who is they?") and another for emotions ("how do they feel?"). These processes are informed by the static and dynamic features of the face, respectively (Bruce & Young, 1986; Haxby et al., 2000).

Similar to the workings of human brains, automatic systems for FER are largely independent from systems for identity recognition, although they typically share a stage of face detection. Propelled by the prospect of dependable and affordable applications across various contexts, the advancement of FER has catalyzed a substantial and continually expanding market.

Confronted with a 2D or 3D image or video portraying a facial expression, a system for FER assigns it to some emotion category. This process initiates with an image preprocessing stage: a face detection algorithm isolates the facial region, normalizes its orientation, eliminates irrelevant details, and standardizes pertinent information to facilitate feature extraction. Then, in traditional systems several features are extracted from the image based on proxies such as a pixel's brightness compared to neighboring ones, or the difference in distance between facial landmarks signaling the movement of some facial muscle (e.g. mouth corners, end of eyebrows), which are sometimes interpreted in terms of Ekman's FACS. The final stage involves classifying target images into specific emotion categories by comparing them with the training data (Huang, Chen, Lv, & Wang, 2019).

More recently, a number of FER systems have adopted deep learning algorithms. These systems not only exhibit improved performance but also possess the advantage of reduced reliance on preprocessing and feature extraction. Furthermore, they have demonstrated increased resilience in noisy (and hence more realistic) environments.

FER are facing many technical challenges. For example, they require training datasets that strike a balance between diversity and ecological validity while avoiding excessive computational demands (Huang et al., 2019; Li & Deng, 2020).

Dupré, Krumhuber, Küster, and McKeown (2020) conducted a study comparing the accuracy of eight commercially available classifiers with that of human observers. This comparison was based on dynamic facial expressions, encompassing both spontaneous and posed expressions. The accuracy of FER systems ranged from 48% to 62%, whereas human observers achieved an accuracy of 75%. Moreover, FER systems displayed notably lower accuracy in identifying spontaneous expressions. In contrast to human judgments, FER systems appear to be more reliant on the conformity to prototypical expressions. This entails adhering to the exact action units anticipated by Ekman's theory (Krumhuber, Küster, Namba, & Skora, 2021).

In a separate study, Tcherkassof and Dupré (2021) conducted research that combined human observers' judgments and machines' assessments with self-reports from individuals producing spontaneous, albeit experimentally induced, dynamic expressions. Once again, human observers' judgments were more in line with the self-reports of those displaying the expressions, compared to the judgments made by machines. Nonetheless, even human observers were not without their imperfections.

Addressing the limitations of FER involves more than just technical refinements. To enhance the accuracy of FER and, importantly, to dispel ill-founded expectations (see Fig. 1), several conceptual challenges must also be addressed or at least taken in account while interpreting results. A few notable examples include:

1. Certain facial movements might carry meanings beyond emotions, such as non-emotional connotations or paralinguistic functions. While some scholars, notably Crivelli and Fridlund (2018), argue against considering facial movements as exclusively tied to emotions, it's worth noting that even Ekman himself acknowledged the multifunctional nature of facial movements, including their potential paralinguistic roles (Ekman & Friesen, 1969). FER, constrained by the requirement to assign movements to predefined

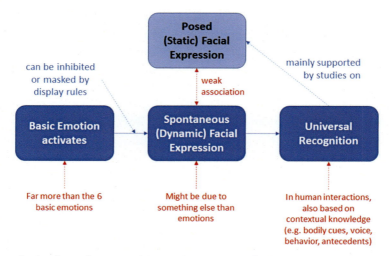

Fig. 1 A visual graph summarizing various stages of emotion recognition (in dark blue) according to Ekman's simplified account, together with possible complications (in light blue) and common criticisms (in red).

emotion categories, faces the risk of misinterpreting a movement executed for other purposes as an emotional signal.
2. In many human interactions, emotions are *underdetermined* by facial movements. Although the inherent prominence of the face might lead us to believe that most emotional information is conveyed through facial cues (Todorov, 2017, chapter 14), some studies reveal that contextual knowledge or bodily cues can alter the emotional interpretation attributed to a given facial expression (see for instance Aviezer et al. 2011; Barrett, Mesquita, & Gendron, 2011). To address this challenge, it becomes necessary to integrate facial movements with information from other communication channels, such as bodily movements or vocal cues.
3. Facial movements can be suppressed or feigned. According to Ekman (1972), this often occurs to conform with *display rules*—social norms dictating the appropriateness of certain emotions in specific situations. Despite Ekman's assertion that traces of spontaneous emotional behavior can still be inferred by detecting the so-called *micro-expressions*, instances of complete expressions that display rules fail to suppress or override, their evidential basis is not especially robust (see Crivelli & Fridlund, 2018). Ironically, most stimuli and databases used in psychological research and FER development are rooted in "feigned" facial expressions—*deliberately* posed expressions (Masson, Cazenave, Trombini, & Batt, 2020). Even expressions referred to

as *spontaneous,* when recreated in controlled lab environments, might lack ecological validity (see Zloteanu & Krumhuber, 2021). While some authors describe software that demonstrates the potential to differentiate between "genuine" and "fake" emotional facial expressions—such as Huynh and Kim (2017) and Wan et al. (2017)—their claims rest largely on databases that do not enable robust, ecologically valid inferences due to their elicitation methods (Ofodile et al., 2017).[3] A more promising approach involves videos from real-life settings (see Cowen et al., 2021; Srinivasan & Martinez, 2018), albeit this comes with a trade-off of less experimental control.

4. Studies of facial movements in real-life settings also highlight a major weak spot of the whole FER enterprise, that is to confine themselves to the "canonical six" emotional categories—namely, anger, disgust, fear, happiness, sadness, surprise. The necessity to transcend this limited taxonomy becomes evident upon considering studies that embrace broader taxonomies or allow for taxonomy emergence through data-driven approaches. A case in point is the ecological study by Cowen et al. (2021), which examines 6 million YouTube videos in emotionally charged contexts (e.g. weddings, funerals). This study identifies reliable connections between 16 facial movement patterns and an equal number of emotion categories (whose affective implications are estimated via the video context, used as a proxy, e.g. weddings and funerals are supposed to be associated with happy and sad feelings, respectively). An even more bottom-up methodology is pursued by Srinivasan and Martinez (2018). Beginning with real-life videos, they pinpointed 8 recurring facial movement patterns within specific cultures and 35 patterns recurring across diverse cultures. They then solicited interpretations of emotional meaning from individuals of various cultures, resulting in a bottom-up taxonomy of facial expressions of emotions that far surpasses the conventional norm.

To sum up, any claimed success of FER warrants careful interpretation. More often than not, such success signifies an ability to identify similarities between facial movements and the prototype movements that Ekman and his followers have traditionally considered as dependable indicators of emotions. While this achievement is impressive within the realm of

---

[3] To elicit respectively "genuine" and "fake" expressions of emotions, Ofodile and colleagues (2017) asked subjects to watch an emotionally suggestive videoclip and then to pose either a congruent expression (for genuine condition) or an incongruent one (for fake one).

machine vision, it's merely an initial step—one that can sometimes be misleading—toward generating authentic and ecologically valid predictions of emotions.

The fact FER systems appear to replicate the human ability to perceive (and potentially misinterpret) emotional expressions through someone's face might inadvertently lead us to underestimate the potential privacy concerns they pose. Given that a neurotypical human can already gauge another person's emotions through biological vision, could it be tempting to dismiss concerns regarding accessing emotions via technologically mediated vision? However, it is worth remembering that, even if FER managed to solve all the epistemic issues that undermine their accuracy, there would still be some stark differences between two cases: in fact, technologically-mediated vision eschews the subject from spatiotemporal constraints, that is being there physically (and most likely offering the same kind of access to facial cues of affective states). Not to mention that it enables the capture, recording, and potentially infinite dissemination of the same emotion-relevant facial cues. Moreover, when subjects are not aware of their facial movements being analyzed by a FER system, they might be less likely to modulate their facial behavior than in face-to-face interactions (see Glazer, 2019). On top on that, the aura of technological and scientific authority surrounding FER systems, fueled by marketing efforts, can lead some inexpert users to place more trust in them than they deserve. Lastly, taking a somewhat radical perspective, we might consider that just because our facial affective cues are *typically accessible* in face-to-face interactions (even when we might wish otherwise), this may not necessarily align with what is *morally desirable*.

3. Decoding emotion from (neurovascular proxies for) brain data

fMRI empowers researchers with the remarkable ability to predict mental states with astonishing accuracy, earning it the honorary label of *"The new mind reader"*, as per the title of a recent book (Poldrack 2018).[4] Through fMRI, researchers can determine the observed object category by a subject (Haxby et al., 2001), forecast simple decisions seconds in advance (Haynes, 2011).

---

[4] For an insightful discussion of the appropriateness of the notion of "mind reading", see Rathkopf et al. (2023).

If paired with recent computational methods for enhancing and constraining the interpretation of signal, fMRI allows to decode visual scenes (Ozcelik & VanRullen, 2023) or the semantic meaning of perceived or imagined language (Tang, LeBel, Jain, & Huth, 2023) with impressive levels of accuracy. And of course, it can be used to predict (or read, decode, infer—there is some lexical variation) the emotion or affective state experienced by a subject. While other neuroimaging techniques like the Electroencephalogram have better temporal resolution, and other techniques like intracortical recordings offer finer-grained spatial resolution, fMRI remains highly popular in neuroscience (especially in affective neuroscience—see Giannakopoulou, Lettieri, Handjaras, Viola, & Cecchetti, in, preparation) due to its robust spatial resolution and non-invasive protocols. In most cases, fMRI tracks the BOLD signal,[5] an endogenous brain signal based on the distinct magnetic properties of oxygenated and deoxygenated hemoglobin. This signal's dynamics correlate with neural activity in a given brain region (Logothetis, 2008).

While colloquial terms sometimes refer to "active brain regions", it's important to emphasize that brain regions are never entirely inactive in a living brain. Instead, they vary in activation (i.e., neuron firing frequency) based on the cognitive task they are engaged in. However, since no region is selectively engaged by a *single* cognitive task (Anderson, Kinnison, & Pessoa, 2013), inferring a specific mental state based on regional activity is often sanctioned as the reverse inference fallacy (Poldrack, 2006). However, in the last two decades, prediction accuracy has significantly improved as researchers have shifted focus from the activity of individual regions to the pattern of activity across multiple voxels,[6] utilizing a technique called MultiVoxel Pattern Analysis (MVPA; Haxby et al., 2014). In MVPA, each of the n voxels' activity is represented as a point in an n-dimensional space. Often, a classifier (a machine learning algorithm) is trained to associate these patterns with corresponding categories of mental states entertained by the subject, such as specific categories of visual stimuli they perceive (e.g. houses versus cats) or emotions (e.g. anger versus happiness). Following successful training, the classifier can infer the category of the mental process most likely entertained by the subject based on the corresponding neural patterns. Since human brains all have different sizes and shapes, several alignment techniques have been developed to predict mental states of one subject using training data from other subjects' brains (e.g. Haxby et al., 2011).

---

[5] BOLD stands for *Blood Oxygenation Level Dependent*.
[6] Voxels are tridimensional pixels in which the neuroimage is parsed.

However, alignment techniques tailored for specific purposes may not generalize to other purposes (Ward, 2022) and might not account for idiosyncratic mind-brain organizations (Viola, 2021).

Another crucial point to highlight is that our interpretation of brain signals is inherently limited by our cognitive ontology (Poldrack, 2010), which refers to how we conceptualize cognitive tasks in the context of psychological theories, or by the categories we employ to train classifiers.

While universalistic perspectives on the neural organization of emotion (e.g., Ekman, 1992; see also Adolphs, 2017) view the challenge of individual differences as a complicating factor in correlating specific emotions with neural activity patterns, certain theories of emotion, particularly within the constructionist framework (e.g., Barrett, 2017), suggest that individual differences might also extend to which emotions are incorporated into one's cognitive ontology. In this view, extreme variations in an individual's ontology of emotion could potentially hinder the possibility of establishing generalizable mental-neural correlations. Even in the scenario where a shared set of emotions exists across all individuals, an essential question persists: Which categories should be encompassed within this set?

In many fMRI experiments, neuroscientists assumed a universal ontology of at least some emotions. Which ones? Of course, those from Ekman's taxonomy. Indeed, several experiments looked for the neural correlates of experienced or perceived anger, disgust, fear, joy, and sadness (Giannakopoulou et al., in preparation)— whereas surprise was poorly studied. Even before the surge of MVPA, several meta-analyzes, taking into account hundreds of studies about Ekman's basic emotions (except surprise) firmly demonstrated that no emotion category exhibits a one-to-one correspondence with a distinct neural region. Instead, each emotion tends to be consistently associated with distributed networks within the brain (Vytal & Hamann, 2010; Lindquist, Wager, Kober, Bliss-Moreau, & Barrett, 2012; Wager et al., 2015. For a discussion, see Celeghin, Diano, Bagnis, Viola, & Tamietto, 2017).

Some MVPA studies have ventured to predict basic emotions through their corresponding neural patterns (see Kragel & LaBar, 2016). A notable instance includes a widely recognized study by Saarimäki et al. (2016), wherein emotions were elicited through two distinct methods (brief video clips and self-induction). Impressively, this study achieved above-chance prediction accuracy not only across different elicitation modalities but also across individual subjects. In essence, classifiers trained with neural correlates of self-induced emotions displayed the ability to predict the emotional response triggered by short video clips, based on corresponding neural patterns, and vice versa. Even more remarkably, the classifier demonstrated the capacity to forecast an individual's emotional state using their neural data, even when trained exclusively on data from other subjects.

Prima facie, such outcomes appear to lend support to Ekman's taxonomy of fundamental emotions from a neural perspective, as he predicted that "There must be unique physiological *patterns* for each emotion, and these [central nervous system] *patterns* should be specific to these emotions not found in other mental activity" (Ekman., 1992, p. 182. Emphasis mine). However, the situation is more intricate. A few years later, the same neuroscience laboratory conducted another fMRI study involving eight "non-basic" emotion categories and discovered relatively consistent neural mappings for many of these categories as well (Saarimäki et al., 2018). Moreover, recent investigations have identified consistent neural underpinnings for a multitude of emotions across subjects, ranging up to 36 distinct categories (Horikawa, Cowen, Keltner, & Kamitani, 2020; Koide-Majima, Nakai, & Nishimoto, 2020).

While Ekman's emotion categories enjoy popularity in affective neuroscience, they are not the sole game in town. Indeed, a diverse array of fMRI experiments adopts alternative cognitive ontologies of emotion. Some studies take exploratory, bottom-up approaches, linking neural responses to self-reported qualities of stimuli (e.g., Lettieri et al., 2019). But the prevalent rival to Ekman's ontology derives from the two-dimensional constructs often considered key aspects of affective experience—*valence* (ranging from pleasure to displeasure) and *arousal* (from deactivated to activated) (Russell, 2003).

Numerous MVPA investigations have successfully predicted the valence of specific stimuli, uncovering both modality-specific and supramodal brain regions responsive to arousal and valence (Baucom, Wedell, Wang, Blitzer, & Shinkareva, 2012; Chikazoe, Lee, Kriegeskorte, & Anderson, 2014). However, in contrast to the robust neural predictions derived from basic emotions, the neural underpinnings of valence and arousal levels have demonstrated weaker correlations with underlying neural activity (Kragel & LaBar, 2015; Wager et al., 2015). This suggests that the psychometric solidity of their phenomenological manifestations might not extend as seamlessly to the neural level (Barrett & Bliss-Moreau, 2009).

Recent research trends even propose that valence might not be uniformly represented across the brain along a single bipolar positive-to-negative continuum. Certain brain areas seem to adhere to such representation, whereas others display sensitivity exclusively to positive or negative stimuli, or to arousal (Mattek, Burr, Shin, Whicker, & Kim, 2020; Vaccaro, Kaplan, & Damasio, 2020). This prompts consideration of the idea that the brain's depiction of affective states may not adhere to a singular and straightforward ontology, with categories closely resembling folk emotion concepts (Scarantino, 2012). Instead, the brain might employ a multitude of diverse constructs operating at various layers of its processing.

Yet other challenges extend beyond the reliance on Ekman's insufficiently validated emotional framework and encompass the fundamental aspects of fMRI protocols themselves. Given how unlikely it is to elicit authentic emotional experiences from subjects within an MRI scanner, many studies gravitate toward investigating the perception of affective stimuli instead. While these pursuits share neural substrates to some degree (Gallese & Caruana, 2016), the processes of perceiving and experiencing affective states remain partially distinct. Transitioning from one to the other involves an inferential leap that introduces theoretical vulnerabilities.

These concerns can be partly alleviated by employing realistic stimuli that can be readily observed within the confines of an MRI scanner, such as movies (Saarimäki, 2021). Alternatively, researchers might adopt different techniques with a more causally direct approach, such as stimulating specific brain regions (Caruana et al., 2015). Such approaches could help mitigate the limitations posed by the inherent constraints of fMRI protocols and offer new avenues for investigating the neural correlates of affective phenomena.

Despite its numerous limitations, it is reasonable to acknowledge that fMRI provides a relatively direct means of accessing what most scholars consider to be the core of affective states. Notably, studies involving subliminal stimuli (Prochnow et al., 2013) or individuals with pathological conditions preventing conscious vision (Burra, Hervais-Adelman, Celeghin, De Gelder, & Pegna, 2019; Vuilleumier et al., 2002) have successfully identified neural correlates of emotional facial expression perception even in the absence of awareness. This suggests the intriguing possibility that fMRI scanners might sometimes yield an understanding of a subject's affective states that is deeper than the subject's own understanding. An important ethical implication of this notion is that it would be challenging to deceive the scanner regarding our genuine emotional responses to stimuli, even though our own conscious interpretation of the nuances in our neural activity might remain less precise.

That being said, the ethical worries about the power of these "mind readers" should be mitigated by considering their heavy logistic constraints.

Letting aside the burdensome cost of buying and maintaining an up-to-date MR scanner, that makes their large-scale use almost impossible, the fact that these scanners require subjects to lie down and stay still makes them ill-suited for answering many ecologically interesting questions and unavailable for many real-life contexts.

4. Summary and open problems

In this chapter, I suggest that, much like the (no longer optimal) design of QWERTY keyboards, which is rooted in their ancestor devices (namely, writing machines), the (not always optimal) machinery governing our emotions is a phylogenetic inheritance from our ancestors, according to a Darwinian framework. Similarly, our current theories of emotion inherit some (no longer optimal) assumptions advocated by an influential psychologist who followed in Darwin's footsteps, namely Paul Ekman. Two prominent features of Ekman's legacy are the centrality of (a stereotyped set of posed) facial movements, often assumed to express a specific universal emotion, and the consolidation of a taxonomy of six "canonical" emotion categories that are supposed to neatly map with these movements.

Despite heavy criticism in affective science, these theoretical features have exerted considerable influence on the development of two popular techniques that try to decode emotions from facial movements and neurovascular metabolism, namely FER and fMRI, respectively. The scope of both techniques seems constrained by the fact that emotions may be more numerous than the canonical six Ekman envisaged, and their facial and neural correlates may be more varied than what Ekman universalistic framework suggests.

Perhaps, in the not-so-distant future, affective computing will loosen its bonds to Ekman's cumbersome legacy and ameliorate its decoding accuracy. Would that impact the ethical issues posed by FER and fMRI? Perhaps it will, albeit only indirectly. Intuitively, if someone tries to discern my affective state against my will, I would not find it more or less violated based on whether they decode them correctly or not. Instead, my discomfort would probably scale up linearly with the degree of confidence they have about how I feel. In that respect, what matters is not so much the actual accuracy of the techniques but rather their perceived accuracy.

In any case, which technique should we fear the most? In optimal conditions, fMRI scanners could provide more accurate decoding and cannot be deceived as easily as FER. Yet, these optimal conditions require having a million-dollar scanner running with up-to-date analysis protocols operating for hours of training, whose efficacy depends on the collaboration of the scanned subject. In contrast, although less reliable, FER is way cheaper and can be pervasively implemented unbeknownst to subjects. This simple consideration suggests that ethical debates on mental privacy should not confine themselves to "neural data" but scrutinize also (and perhaps, primarily) mental data collected from facial signals and other behavioral sources (e.g., Ienca & Malgieri, 2022).

Acknowledgments

This chapter is largely based on discussions with Marcello Ienca. The author is also grateful to Pia Campeggiani for her insightful comments and to Georg Starke & Marcello Ienca for their great editorial work. ChatGPT 3.5 has been employed for proofreading.

References

Adolphs, R. (2017). How should neuroscience study emotions? By distinguishing emotion states, concepts, and experiences. *Social Cognitive and Affective Neuroscience, 12*(1), 24–31.

Anderson, M. L., Kinnison, J., & Pessoa, L. (2013). Describing functional diversity of brain regions and brain networks. *Neuroimage, 73*, 50–58.

Aviezer, H., Bentin, S., Dudarev, V., & Hassin, R. R. (2011). The automaticity of emotional face-context integration. *Emotion (Washington, D. C.), 11*(6), 1406.

Barrett, L. F. (2017). The theory of constructed emotion: An active inference account of interoception and categorization. *Social Cognitive and Affective Neuroscience, 12*(1), 1–23.

Barrett, L. F., & Bliss-Moreau, E. (2009). Affect as a psychological primitive. *Advances in Experimental Social Psychology, 41*, 167–218.

Barrett, L. F., Mesquita, B., & Gendron, M. (2011). Context in emotion perception. *Current Directions in Psychological Science, 20*(5), 286–290.

Barrett, L. F., Adolphs, R., Marsella, S., Martinez, A. M., & Pollak, S. D. (2019). Emotional expressions reconsidered: Challenges to inferring emotion from human facial movements. *Psychological science in the public interest, 20*(1), 1–68.

Baucom, L. B., Wedell, D. H., Wang, J., Blitzer, D. N., & Shinkareva, S. V. (2012). Decoding the neural representation of affective states. *Neuroimage, 59*(1), 718–727.

Benda, M. S., & Scherf, K. S. (2020). The Complex Emotion Expression Database: A validated stimulus set of trained actors. *PLoS One, 15*(2), e0228248.

Bruce, V., & Young, A. (1986). Understanding face recognition. *British Journal of Psychology, 77*(3), 305–327.

Buiatti, M., Di Giorgio, E., Piazza, M., Polloni, C., Menna, G., Taddei, F., & Vallortigara, G. (2019). Cortical route for facelike pattern processing in human newborns. *Proceedings of the National Academy of Sciences, 116*(10), 4625–4630.

Burra, N., Hervais-Adelman, A., Celeghin, A., De Gelder, B., & Pegna, A. J. (2019). Affective blindsight relies on low spatial frequencies. *Neuropsychologia, 128*, 44–49.

Caruana, F., Avanzini, P., Gozzo, F., Francione, S., Cardinale, F., & Rizzolatti, G. (2015). Mirth and laughter elicited by electrical stimulation of the human anterior cingulate cortex. *Cortex; A Journal Devoted to the Study of the Nervous System and Behavior, 71*, 323–331.

Celeghin, A., Diano, M., Bagnis, A., Viola, M., & Tamietto, M. (2017). Basic emotions in human neuroscience: Neuroimaging and beyond. *Frontiers in Psychology, 8*, 1432.

Chikazoe, J., Lee, D. H., Kriegeskorte, N., & Anderson, A. K. (2014). Population coding of affect across stimuli, modalities and individuals. *Nature Neuroscience, 17*(8), 1114–1122.

Cohn, J. F., & De la Torre, F. (2015). Automated face analysis for affective computing. In R. A. Calvo, S. D'Mello, J. M. Gratch, & A. Kappas (Eds.). *The Oxford handbook of affective computing* (pp. 131–150). Oxford University Press.

Cowen, A. S., & Keltner, D. (2021). Semantic space theory: A computational approach to emotion. *Trends in Cognitive Sciences, 25*(2), 124–136.

Cowen, A. S., Keltner, D., Schroff, F., Jou, B., Adam, H., & Prasad, G. (2021). Sixteen facial expressions occur in similar contexts worldwide. *Nature, 589*(7841), 251–257.

Crivelli, C., & Fridlund, A. J. (2018). Facial displays are tools for social influence. *Trends in Cognitive Sciences, 22*(5), 388–399.

Darwin, C. (1872). *The expression of emotion in man and animals.* London: Murray.

Dawel, A., Miller, E. J., Horsburgh, A., & Ford, P. (2021). A systematic survey of face stimuli used in psychological research 2000–2020. *Behavior Research Methods,* 1–13.

Dupré, D., Krumhuber, E. G., Küster, D., & McKeown, G. J. (2020). A performance comparison of eight commercially available automatic classifiers for facial affect recognition. *PLoS One, 15*(4), e0231968.

Ekman, P. (1992). An argument for basic emotions. *Cognition & Emotion, 6*(3-4), 169–200.

Ekman, P. (1998). Preface and commentary (1872/1998) In C. Darwin (Ed.). *The expression of the emotions in man and animals.* London: Harper Collins.

Ekman, P., & Friesen, W. V. (1969). The repertoire of nonverbal behavior: Categories origins, usage, and coding. *Semiotica, 1*(1), 49–98.

Ekman, P., & Friesen, W. V. (1971). Constants across cultures in the face and emotion. *Journal of Personality and Social Psychology, 17*(2), 124.

Ekman, P., & Friesen, W. V. (1978). Facial action coding system. *Environmental Psychology & Nonverbal Behavior.*

Ekman, P., Sorenson, E. R., & Friesen, W. V. (1969). Pan-cultural elements in facial displays of emotion. *Science (New York, N. Y.), 164*(3875), 86–88.

Elfenbein, H. A., & Ambady, N. (2002). On the universality and cultural specificity of emotion recognition: A meta-analysis. *Psychological Bulletin, 128*(2), 203.

Ellsworth, P. C. (2014). Basic emotions and the rocks of New Hampshire. *Emotion Review, 6*(1), 21–26.

Fridlund, A. J. (2017). The behavioral ecology view of facial displays, 25 years later. In J. M. F. Dols, & J. A. Russell (Eds.). *The science of facial expression* (pp. 77–92). Oxford University Press.

Gallese, V., & Caruana, F. (2016). Embodied simulation: Beyond the expression/experience dualism of emotions. *Trends in Cognitive Sciences, 20*(6), 397–398.

Giannakopoulou, A., Lettieri, G., Handjaras, G., Viola, M., Cecchetti, L. (in preparation). How has affective neuroscience evolved over the last decades?

Glazer, T. (2019). The social amplification view of facial expression. *Biology & Philosophy, 34*(2), 33.

Griffiths, P. E. (1997). *What emotions really are: The problem of psychological categories.* University of Chicago Press.

Haxby, J. V., Hoffman, E. A., & Gobbini, M. I. (2000). The distributed human neural system for face perception. *Trends in Cognitive Sciences, 4*(6), 223–233.

Haxby, J. V., Connolly, A. C., & Guntupalli, J. S. (2014). Decoding neural representational spaces using multivariate pattern analysis. *Annual Review of Neuroscience, 37*, 435–456.

Haxby, J. V., Gobbini, M. I., Furey, M. L., Ishai, A., Schouten, J. L., & Pietrini, P. (2001). Distributed and overlapping representations of faces and objects in ventral temporal cortex. *Science (New York, N. Y.), 293*(5539), 2425–2430.

Haxby, J. V., Guntupalli, J. S., Connolly, A. C., Halchenko, Y. O., Conroy, B. R., Gobbini, M. I., & Ramadge, P. J. (2011). A common, high-dimensional model of the representational space in human ventral temporal cortex. *Neuron, 72*(2), 404–416.

Haynes, J. D. (2011). Decoding and predicting intentions. *Annals of the New York Academy of Sciences, 1224*(1), 9–21.

Horikawa, T., Cowen, A. S., Keltner, D., & Kamitani, Y. (2020). The neural representation of visually evoked emotion is high-dimensional, categorical, and distributed across transmodal brain regions. *iScience, 23*(5), 101060.

Huang, Y., Chen, F., Lv, S., & Wang, X. (2019). Facial expression recognition: A survey. *Symmetry, 11*(10), 1189.

Huynh, X. P., & Kim, Y. G. (2017). Discrimination between genuine versus fake emotion using long-short term memory with parametric bias and facial landmarks. In *Proceedings of the IEEE international conference on computer vision workshops* (pp. 3065–3072).

Ienca, M., & Malgieri, G. (2022). Mental data protection and the GDPR. *Journal of Law and the Biosciences, 9*(1), lsac006.

Koide-Majima, N., Nakai, T., & Nishimoto, S. (2020). Distinct dimensions of emotion in the human brain and their representation on the cortical surface. *Neuroimage, 222*, 117258.

Kragel, P. A., & LaBar, K. S. (2015). Multivariate neural biomarkers of emotional states are categorically distinct. *Social Cognitive and Affective Neuroscience, 10*(11), 1437–1448.

Kragel, P. A., & LaBar, K. S. (2016). Decoding the nature of emotion in the brain. *Trends in Cognitive Sciences, 20*(6), 444–455.

Krumhuber, E. G., Küster, D., Namba, S., & Skora, L. (2021). Human and machine validation of 14 databases of dynamic facial expressions. *Behavior Research Methods, 53*, 686–701.

Kuhn, T. S. (1970). *The structure of scientific revolutions* (2nd ed.). Chicago University Press.

Lee, D. H., & Anderson, A. K. (2017). Form and function of facial expressive origin. In J. M. Fernandez-Dols, & J. A. Russell (Eds.). *The science of facial expression* (pp. 173–196). Oxford University Press.

Lettieri, G., Handjaras, G., Ricciardi, E., Leo, A., Papale, P., Betta, M., & Cecchetti, L. (2019). Emotionotopy in the human right temporo-parietal cortex. *Nature Communications, 10*(1), 5568.

Li, S., & Deng, W. (2020). Deep facial expression recognition: A survey. *IEEE transactions on affective computing, 13*(3), 1195–1215.

Lindquist, K. A., Wager, T. D., Kober, H., Bliss-Moreau, E., & Barrett, L. F. (2012). The brain basis of emotion: A meta-analytic review. *Behavioral and brain sciences, 35*(3), 121–143.

Logothetis, N. K. (2008). What we can do and what we cannot do with fMRI. *Nature, 453*(7197), 869–878.

Masson, A., Cazenave, G., Trombini, J., & Batt, M. (2020). The current challenges of automatic recognition of facial expressions: A systematic review. *AI Communications, 33*(3-6), 113–138.

Mattek, A. M., Burr, D. A., Shin, J., Whicker, C. L., & Kim, M. J. (2020). Identifying the representational structure of affect using fMRI. *Affective Science, 1*, 42–56.

Miolla, A., Cardaioli, M., & Scarpazza, C. (2022). Padova Emotional Dataset of Facial Expressions (PEDFE): A unique dataset of genuine and posed emotional facial expressions. *Behavior Research Methods*, 1–16.

Ofodile, I., Kulkarni, K., Corneanu, C. A., Escalera, S., Baro, X., Hyniewska, S., ... Anbarjafari, J. (2017). Automatic recognition of deceptive facial expressions of emotion. *arXiv, 04061*.

Ozcelik, F., & VanRullen, R. (2023). Natural scene reconstruction from fMRI signals using generative latent diffusion. *Scientific Reports, 13*(1), 15666.

Panksepp, J. (1998). *Affective neuroscience: The foundations of human and animal emotions*. Oxford University Press.

Plamper, J. (2015). *The history of emotions: An introduction*. Oxford: OUP.

Poldrack, R. (2018). *The new mind readers: What neuroimaging can and cannot reveal about our thoughts*. Princeton University Press.

Poldrack, R. A. (2006). Can cognitive processes be inferred from neuroimaging data? *Trends in Cognitive Sciences, 10*(2), 59–63.

Poldrack, R. A. (2010). Mapping mental function to brain structure: How can cognitive neuroimaging succeed? *Perspectives on psychPoldrack*, 2018.

Prochnow, D., Kossack, H., Brunheim, S., Müller, K., Wittsack, H. J., Markowitsch, H. J., & Seitz, R. J. (2013). Processing of subliminal facial expressions of emotion: A behavioral and fMRI study. *Social Neuroscience, 8*(5), 448–461.

Rathkopf, C., Heinrichs, J. H., & Heinrichs, B. (2023). Can we read minds by imaging brains? *Philosophical Psychology, 36*(2), 221–246.

Russell, J. A. (1994). Is there universal recognition of emotion from facial expression? A review of the cross-cultural studies. *Psychological bulletin, 115*(1), 102–141.

Russell, J. A. (2003). Core affect and the psychological construction of emotion. *Psychological Review, 110*(1), 145.

Saarimäki, H. (2021). Naturalistic stimuli in affective neuroimaging: A review. *Frontiers in Human Neuroscience, 15*, 675068.

Saarimäki, H., Gotsopoulos, A., Jääskeläinen, I. P., Lampinen, J., Vuilleumier, P., Hari, R., & Nummenmaa, L. (2016). Discrete neural signatures of basic emotions. *Cerebral Cortex, 26*(6), 2563–2573.

Saarimäki, H., Ejtehadian, L. F., Glerean, E., Jääskeläinen, I. P., Vuilleumier, P., Sams, M., & Nummenmaa, L. (2018). Distributed affective space represents multiple emotion categories across the human brain. *Social Cognitive and Affective Neuroscience, 13*(5), 471–482.

Scarantino, A. (2012). How to define emotions scientifically. *Emotion Review, 4*(4), 358–368.

Shariff, A. F., & Tracy, J. L. (2011). What are emotion expressions for? *Current Directions in Psychological Science, 20*(6), 395–399.

Srinivasan, R., & Martinez, A. M. (2018). Cross-cultural and cultural-specific production and perception of facial expressions of emotion in the wild. *IEEE Transactions on Affective Computing, 12*(3), 707–721.

Stark, L., & Hoey, J. (2021, March). The ethics of emotion in artificial intelligence systems. In *Proceedings of the 2021 ACM conference on fairness, accountability, and transparency* (pp. 782–793).

Tang, J., LeBel, A., Jain, S., & Huth, A. G. (2023). Semantic reconstruction of continuous language from non-invasive brain recordings. *Nature Neuroscience*, 1–9.

Tcherkassof, A., & Dupré, D. (2021). The emotion–facial expression link: Evidence from human and automatic expression recognition. *Psychological Research, 85*(8), 2954–2969.

Todorov, A. (2017). *Face value: The irresistible influence of first impressions*. Princeton University Press.

Tooby, J., & Cosmides, L. (1990). The past explains the present: Emotional adaptations and the structure of ancestral environments. *Ethology and Sociobiology, 11*(4-5), 375–424.

Vaccaro, A. G., Kaplan, J. T., & Damasio, A. (2020). Bittersweet: The neuroscience of ambivalent affect. *Perspectives on Psychological Science, 15*(5), 1187–1199.

Viola, M. (2021). Beyond the platonic brain: Facing the challenge of individual differences in function-structure mapping. *Synthese, 199*(1-2), 2129–2155.

Vuilleumier, P., Armony, J. L., Clarke, K., Husain, M., Driver, J., & Dolan, R. J. (2002). Neural response to emotional faces with and without awareness: Event-related fMRI in a parietal patient with visual extinction and spatial neglect. *Neuropsychologia, 40*(12), 2156–2166.

Vytal, K., & Hamann, S. (2010). Neuroimaging support for discrete neural correlates of basic emotions: A voxel-based meta-analysis. *Journal of Cognitive Neuroscience, 22*(12), 2864–2885.

Wager, T. D., Kang, J., Johnson, T. D., Nichols, T. E., Satpute, A. B., & Barrett, L. F. (2015). A Bayesian model of category-specific emotional brain responses. *PLoS Computational Biology, 11*(4), e1004066.

Wan, J., Escalera, S., Anbarjafari, G., Jair Escalante, H., Baró, X., Guyon, I., & Xie, Y. (2017). Results and analysis of chalearn lap multi-modal isolated and continuous gesture recognition, and real versus fake expressed emotions challenges. In *Proceedings of the IEEE international conference on computer vision workshops* (pp. 3189–3197).

Ward, Z. B. (2022). Registration pluralism and the cartographic approach to data aggregation across brains. *The British Journal for the Philosophy of Science, 73*(1).

Zloteanu, M., & Krumhuber, E. G. (2021). Expression authenticity: The role of genuine and deliberate displays in emotion perception. *Frontiers in Psychology, 11*, 611248.

SECTION 3

Finding common ground

CHAPTER EIGHT

Algorithmic regulation: A compatible framework for AI and DTC neurotechnologies

Lucille Nalbach Tournas[a,*] and Walter G. Johnson[b]
[a]School of Life Sciences, Arizona State University, Tempe, AZ, United States
[b]School of Regulation and Global Governance (RegNet), The Australian National University, Canberra ACT, Australia
*Corresponding author. e-mail address: lucille.tournas@asu.edu

Contents

1. Introduction	144
2. Direct to consumer neurotechnologies	145
3. Algorithmic regulation	147
4. Applying algorithmic regulation to neurotechnology	149
4.1 Privacy and surveillance	150
4.2 Autonomy and decision-making	152
4.3 Personality and identity	155
5. Conclusion	156
References	157
Further reading	159

Abstract

As direct-to-consumer neurotechnologies begin to enter markets around the world with much promise and concern, conversations remain ongoing around how AI products and services may affect societies and shape behavior. These debates in AI offer productive and complementary concepts for scholars addressing common governance questions in neurotechnology such as data protection, identity, autonomy, and rights. In this chapter, we argue the concept of "algorithmic regulation"—advanced by Yeung in 2018—can offer fresh perspectives to neuroethics and studies of neurotechnology governance. Algorithmic regulation as a lens primarily examines how digital technologies "regulate" individuals and societies by shaping, enabling, and constraining behavior, preferences, and values.

Especially as neurotechnologies become increasingly data- and software-driven, while DTC products and data begin to operate in a global marketplace, algorithmic regulation offers a good fit for opening analysis around not only ethical and regulatory issues around neurotechnologies, but also political economic questions. We connect and compare neuroethical conversations around, for example, personality and identity to algorithmic regulation scholarship examining how existing digital and big data-based tools already alter or erode autonomy and enact new forms of surveillance. In turn, neuroethical studies can contribute back to algorithmic regulation by guiding analysts to consider further how the use of highly sensitive data and intimate digital-biological connections can raise new or modified political and regulatory questions for how technology regulates people and populations.

1. Introduction

Direct-to-consumer (DTC) neurotechnologies have begun to enter markets around the world with much promise and excitement (Ienca, Haselager, & Emanuel, 2018; UNESCO, 2023). These products generally aim to offer new tools to consumers for wellness, entertainment, or workplace intended uses (and, potentially, medical applications to patients), including increasing user focus, aiding in meditation, uses for gaming, and offering the potential of thought-to-text applications. While most of these products primarily act by recording data directly or indirectly from the human nervous system, a smaller subset aims to modulate brain activity through, for example, transcranial electrical stimulation. Accordingly, DTC neurotechnologies also raise several ethical concerns characterized by neuroethicists and national and international organizations, ranging from safety and data protection to more abstract concerns around personality and autonomy (International Bioethics Committee [IBC] Farahany, 2019; Goering et al. 2021; International Bioethics Committee, 2021; Rommelfanger et al. 2018; Wexler & Reiner, 2019).

These debates around the ethics of DTC neurotechnologies occur against the backdrop of ongoing conversations around how artificial intelligence (AI) products and services may affect societies and shape individual behavior. Such discussions in AI have begun to move beyond the ethical realm to include perspectives from a wide variety of other disciplines, including law, surveillance studies, political economy, and regulatory studies. Both fields face common questions around privacy, autonomy, and rights, yet discourse around neurotechnologies has largely addressed these questions from ethical perspectives. Especially given the role of AI in powering and driving many emerging neurotechnologies, the multidisciplinary debates and insights around AI governance can offer productive and complementary concepts for ethicists and other scholars addressing issues around DTC neurotechnology.

In this chapter, we argue the concept of "algorithmic regulation"— advanced especially by Karen Yeung in 2018 by building on work from regulatory studies, law and technology, and science and technology studies—can offer fresh perspective to neuroethics and studies of neuro-technology governance. Algorithmic regulation as a lens examines how digital technologies themselves, and those who deploy them, can broadly "regulate" individuals and societies by shaping, enabling, and constraining behavior, preferences, and values (Bellanova & De Goede, 2022; Ulbricht & Yeung, 2022; Yeung, 2018). Especially as neurotechnologies become increasingly data- and software-driven, while DTC products and data begin to operate in global marketplaces, algorithmic regulation offers a good fit for opening new lines of analysis around not only ethical and regulatory issues around neurotechnologies, but also political economic questions.

The chapter begins by describing the state of DTC neurotechnologies to ground the analysis, before reviewing literature on and underpinning algorithmic regulation. We then connect and compare neuroethical conversations around identity and autonomy to concepts and insights arising from the algorithmic regulation scholarship. This and related work illustrate how existing digital and big data-based tools already alter or erode autonomy and enact new forms of surveillance, which conversations around emerging neurotechnologies can benefit from by becoming more grounded in these existing, empirical conversations. In turn, we consider how neuroethical studies can contribute back to algorithmic regulation by guiding analysts to consider further how the use of highly sensitive data and intimate digital-biological connections can raise new or modified political and regulatory questions for how technology regulates people and populations. While this discussion can only offer a starting point for further inquiry, the chapter considers new opportunities for dialog between AI ethics and neuroethics by exploring how insights from related disciplines can bridge and enrich these conversations.

2. Direct to consumer neurotechnologies

While the term neurotechnology can be difficult to pin down, debate around DTC neurotechnologies generally covers products—devices in particular—that directly or indirectly interact with the human nervous system and are marketed to consumers instead of being intended for clinical or research uses (Ienca et al., 2018; Wexler & Reiner, 2019). These might include wearable brain-computer interfaces (BCI), which enable communication between the human brain and an external computing device, often through portable devices that may use techniques such as electroencephalography (EEG). Firms developing these products may advertise or claim varying wellness benefits or opportunities for entertainment or gaming. These might include smaller firms such as Interaxon or Neurosky and products such as headsets claiming to monitor and improve a consumer's meditation and attention or use brain signals as the input for interacting with a gaming system (Neurosky, 2023). Several large technology companies have also made investments or acquisitions in the DTC neurotechnology space, such as Meta, Snap, or Apple, presenting the possibility of these larger firms offering DTC neurotechnology products to a wide range of potential consumers in the future.

These DTC neurotechnology products tend to come in the form of a wearable device such as a headset, but also generally are powered by a software component, either onboard or off of the device itself. Especially in newer and more sophisticated products, software elements may include systems for collecting, sorting, and processing data—either direct neural data or data that indirectly speaks to neural functioning—and may have algorithms that can alter the functioning of the device in response to certain inputs. These algorithms are often not simple and may rely on complex AI techniques to process, learn, or extrapolate from available data in order to better perform a task or increase product performance (LeCun, Bengio, & Hinton, 2015; Li & Zhao, 2014). Therefore, the convergence of AI and neuroscience in BCI applications could offer novel ways to understand brain function and use such insight for commercial products.

The global market for these devices is growing at significant rates. According to a July 2023 report by the United Nations Educational, Scientific and Cultural Organization (UNESCO, 2023), both public and private investment in research and development towards these technologies are increasing rapidly; with the total neurotechnology market projected to reach $24.2 billion by 2027 (UNESCO, 2023). While these projections include medical devices as well as consumer-facing neurotechnologies, DTC products will likely have lower regulatory burdens in many jurisdictions (International Bioethics Committee, 2021; Wexler & Reiner, 2019). Lower barriers to entry will likely translate to faster and less expensive opportunities for DTC products to enter marketplaces around the world, raising the need to grapple with their ethical as well as regulatory implications.

With this in mind, better understanding the scope and scale of power and unique concerns of the algorithms used in DTC neurotechnology products, such as BCIs, may offer insights valuable for developing regulatory guidance at various levels of government. As digital technologies have become more central to the global economy in the past two decades, algorithms and algorithmic decision-making have become foundational in not only the technology they underlie, but in shaping social processes and behaviors (Cohen, 2019; Zuboff, 2019). Neurotechnology devices have the potential to collect particularly sensitive personal data or influence behavior through new means by more directly interacting with the brain or mind than other digital technologies, raising new or bolstered concerns here. In the following section, we introduce a framework from the regulatory literature—algorithmic regulation—that can assist in conceptualizing how these processes fit within already complex systems of governance. This concept will then be applied to DTC BCIs specifically.

3. Algorithmic regulation

While a common understanding of "regulation" is often top-down "command and control," which is represented by state regulation through the use of hard law and credible (even criminal) sanctions by state actors, regulatory scholars have increasingly challenged and questioned this understanding of regulation, particularly given the contemporary shape of dynamic global marketplaces. More recent developments in regulatory theory conceptualize regulation as untethered from the exclusivity of the state, fragmented between multiple types of state and nonstate actors with various capacities, values, and roles (Black, 2002; Braithwaite and Drahos, 2000; Ford, 2017; Koop and Lodge, 2017). Regulation here becomes a more complex phenomenon, where state, private, and civil society actors are all empowered to regulate themselves and one another through various channels and mechanisms, including through more subtle influences and discursive efforts as well as top-down rulemaking and enforcement (Braithwaite, 2008; Burris, Drahos, & Shearing, 2005).

Scholars of law and technology and science and technology studies have taken these insights in regulation further by arguing technology itself can be a tool of regulation, or even an autonomous regulatory actor. Drawing in part on insights from early studies of the internet proclaiming "code is law," this body of work illustrates how various technologies can contribute to structuring society by empowering or constraining human behavior and autonomy (Aneesh, 2009; Brownsword, 2005; Lessig, 1999). More recent work continues to expand on this notion of techno-regulation by exploring, for example, how private actors can remotely control the functionalities of consumer products that have an internet of things connection, granting private actors new ways to reach into the lives of consumers (Tusikov, 2019). Further, technologies and their development can contribute to processes of policy change alongside framing discourses, politics, and policy alternatives (Goyal, Howlett, & Taeihagh, 2021). This vision of (techno-)regulation arguably better aligns with the complex society in which it lives and perhaps is better understood using social theories such as Actor-Network Theory, developed especially by Latour (2005; see Sayes, 2014), in which human actors and non-human "actants" collectively occupy networks and may equally influence behavior across these networks.

Within this broader framework of techno-regulation, "algorithmic regulation" has recently become a notable concept and line of scholarly inquiry. In a seminal 2018 article, Yeung defines this phenomenon as "regulatory governance systems that utilize algorithmic decision making," meaning "algorithmically generated knowledge systems to execute or inform decisions, which can vary widely in simplicity and sophistication" (2018, p. 507). Yeung continues to argue and explore how algorithms can provide all of the basic functions often associated with regulation—(1) rulemaking, (2) monitoring, and (3) enforcement (see Black, 2002)—and how algorithms can do so in both *ex ante* and *ex post* formats or through both direct decisions and providing recommendations to human actors. Further, more sophisticated algorithms that can continuously learn and update their internal parameters can exercise a degree of autonomy and responsiveness by adjusting its regulatory functions over time and as more data is collected and processed.

This framework of algorithmic regulation can be expanded further to address and analyze a broader set of regulatory phenomena, including more subtle and pervasive forms of power (Ulbricht & Yeung, 2022). For instance, algorithms can create environments where regulated actors realistically have limited autonomy or capacity to resist unfair regulation, for instance as how Uber has used digital means to engage in pervasive and unilateral monitoring and sanctioning of drivers (Eyert, Irgmaier, & Ulbricht, 2022). Both state and nonstate actors may seek to deploy these tools of algorithmic regulation and the goals of these interventions may or may not be determined in advance (Bellanova & De Goede, 2022). Data generally and digital tools like algorithms and platform infrastructure have become particularly prominent in the global economy (Cohen, 2019; Sell, 2022), raising the political economic stakes of algorithmic regulation.

Ethically, Yeung (2019) identifies several points of departure for concerns around algorithmic regulation. These include not only potentially poor or biased outcomes from algorithmic decision-making, but also broader social issues around the (un)democratic nature of algorithmic decision-making and the potential of algorithms to subtly shape individual preferences in ways that can create political and economic winners and losers. This is consistent with law and technology scholars who note the regulatory shift to technological management in place of rules of law (Brownsword, 2019). In closely related work on "hypernudge," Yeung (2017) posits that algorithms powered by big data can create substantial "nudges" to influence behavior in a personalized, predictable, dynamic, and opaque way. Zuboff goes further, suggesting the smart algorithm–data pipeline "unilaterally claims human experience as free raw material for translation into behavioral data [which] are declared as a proprietary behavioral surplus ... and fabricated into prediction products that anticipate what you will do now, soon, and later" (Zuboff, 2019, p.14).

4. Applying algorithmic regulation to neurotechnology

DTC neurotechnologies will not only connect the brain's electrical activity to a computing system in order to record, analyze, and translate them to elicit a specific command, but could also be extensions of algorithmic and data relationships into the mind itself. Therefore, many concerns identified and explored in the algorithmic regulatory literature may provide valuable insights to scholars examining ethical and policy issues for BCIs. Many of the neuroethical concerns surrounding BCIs, including threats to personhood and identity as well as autonomy and decision making, are similar to debates occurring around algorithms already. Concerns around AI and algorithms already explore issues around changes in how individuals, groups, and organizations define their identity and view themselves (Whitley, Gal, & Kjaergaard, 2014). Others express concerns around how personalized algorithmic targeting services impinge on autonomy by seeking to keep users engaged with products and predicting, steering, or limiting behavioral choices while iteratively collecting ever-increasing amounts of data (Yeung, 2017; Zuboff, 2019).

Viewing related neuroethical issues through the lens of algorithmic regulation may allow for a slightly different and potentially productive analysis, shifting attention towards foundational concerns created by the relationships between algorithms and individuals or societies. This section explores the neuroethical topics of privacy and surveillance, autonomy and decision making, and personality and identity to explore how algorithmic regulation may offer a practical companion to neuroethics, one that helps address the fundamental issue of human agency that neuroethics assumes. These topics have been selected to demonstrate the considerable potential overlap between topics already explored in the neuroethical literature and the analytic framework offered by algorithmic regulation.

4.1 Privacy and surveillance

One common issue in neuroethical conversations comes from privacy concerns, especially around the data collected by neurotechnologies such as BCIs sold to consumers in nonclinical settings. The consumer space offers particular concerns, given that health data in jurisdictions such as the United States receives special legal protection, but data collected outside of a medical context—even if it speaks to an individual's or population health—often falls outside of these legal protections (Wexler & Reiner, 2019). Additionally, the neuroethics literature argues in several ways that neural data is uniquely sensitive or vulnerable, particularly given the kinds of insights that future advances in technology may enable to be decoded from EEG and other types of data collected (IBC, 2021; Goering et al., 2021). By emphasizing the unique nature of the data collected, neuroethical conversations—particularly from Western perspectives—often raise concerns about BCIs being able to detect sensitive thoughts or identities and the threats to personal wellbeing, dignity, or safety that may flow from such intrusions (Farahany, 2023; Ienca & Malgieri, 2022).[1] Recent technical advances, such as a 2023 preprint claiming to be able to distinguish between heterosexual and queer men using only resting EEG data (Ziogas et al., 2023) buttress these ethical concerns. Such concerns apply not only in the consumer space, but also in the workplace, classroom, entertainment, and other nonclinical settings (see eg. Farahany, 2023).

These concerns around privacy, largely conceptualized as harms or risks for individuals, can also be seen through the lens of algorithmic regulation.

---

[1] To a lesser extent, some literature considers different cultural notions of privacy as well (Rommelfanger et al., 2019; Herrera-Ferrá et al., 2022).

This and related literatures (e.g., science and technology studies, political economy) critique the collection and processing of data as well, but often from a more structural and population level. This is consistent with more recent legal scholarship treating privacy and data protection as systemic issues with consequences at the group level distinguishable from harms to individuals (Citron, 2023; Cohen, 2019). A primary area of concern here becomes how algorithms—autonomously or by those who design and deploy them, or both—will act to "regulate" individuals and societies through the act of collecting and processing population-level data. Regulatory theory provides that a core feature of regulation is monitoring or information gathering, as this enables further discourse and decision-making about the adequacy of current rules (and their interpretation) and enforcement approaches (Black, 2002). From this perspective, data collection and processing (bringing associated privacy concerns) represent a core regulatory function of algorithms (Yeung, 2018).

In the context of DTC neurotechnologies, collecting particularly sensitive types of neural data may raise the stakes of this type of regulation by monitoring. The potential, or even the fear, that neural data may reveal sensitive information such as personal desires, intentions, or even sexual orientation or gender identity, may create new or altered power dynamics between those who collect data and data subjects, including chilling effects and the potential for privacy harms such as blackmail (see Citron & Solove, 2021). Algorithmic systems can also collect data on whether and how individuals and groups respond to algorithmic decision-making, creating recursive cycles of monitoring, standard setting, and enforcement (Yeung, 2017). This may enable deeper and more nuanced monitoring of neural data by continuously updating and personalizing models which can then be used for further decision-making and influencing of individuals or groups based on potentially sensitive information derived from neural data.

Further, algorithmic monitoring of human actors can empower the state or private actors who use them to surveil populations and (re)impose power dynamics, such as on workers in precarious work environments (Bellanova & De Goede, 2022; Eyert et al., 2022; Henne, 2019). These observations on how actors can use algorithms to regulate other individuals or groups can help strengthen existing arguments in the neuroethics and neurolaw literatures by making tighter connections between ethical concerns around privacy, the types of harms data collection and processing can lead to, and response options. For example, Farahany (2023) supports her argument for a right to cognitive liberty in part by using a provocative vignette depicting workplace neural data collection and the distress this surveillance could place on workers. An algorithmic regulatory lens can be used to redescribe this vignette as a scenario where employers use BCIs to monitor and surveil workers—reifying or exacerbating existing power dynamics in the workplace—for the purpose of later setting private rules about workplace behavior (or thought) and enforcing those standards. In one portion of the vignette, a worker is concerned that the BCI she wears may enable her employer to identify and sanction her intimate thoughts about a colleague (Farahany 2023, p. 6–7). This illustrates the regulatory power of the algorithm embedded in the BCI, as even the fear that the employer could detect a type of thought through data collection contributes to constructing a rule (potentially self-imposed) that intimate thoughts are out-of-bounds in the workplace (see Haggerty & Ericson, 2000).

Revealing the ways that more powerful social actors may use neurotechnologies to surveil and exert power over more marginalized groups can enable a richer conversation around the appropriate response options to this type of private, algorithmic regulation–where novel human rights may be one option to consider. It may also help show why some response options may not be appropriate. For instance, merely adding a human-in-the-loop to this vignette of workplace neural data collection would likely be ineffective—and potentially lack legitimacy (see Martin & Waldman, 2023)—as having humans monitor the data collected and decisions made around that data could contribute to the power dynamics at play in the scenario.

4.2 Autonomy and decision-making

Another area of concern in the neuroethical literature and policy reports comes from the potential for neurotechnologies to impinge on autonomy; again, often conceptualized as the autonomy of individuals (IBC, 2021; Farahany, 2019; Goering et al. 2021; Rommelfanger et al., 2019). Particular concern here is typically given to neurotechnologies which can directly (often electrically) stimulate the brain, raising issues around altering neural pathways and thought patterns. These raise notable issues about informed consent and direct manipulation of individuals through neurotechnological means, as well as broader concerns around the continuity or integrity of personality or identity. In the DTC setting, these neurotechnological devices often receive less oversight than medical products (for now), even when providing transcranial stimulation through various techniques (Wexler & Reiner, 2019).

Algorithmic regulation provides a framework to reinterpret these concerns through a more systemic and group-level lens, potentially revealing new types or different facets of concerns. As opposed to direct infringements on autonomy by electrically altering neural pathways, this paradigm suggests that even BCIs which merely collect neural data can already play a role in influencing behavior and limiting autonomy. The example of workplace surveillance from Farahany (2023) already illustrates this point at the individual level—by using data collection as a technique for surveillance, employers already have the potential to limit autonomy and alter decision-making for workers subject to monitoring. This suggestion is consistent with literature on algorithmic regulation, which has found that platform firms have used data collection to track and influence the behavior of workers (Eyert et al., 2022).

Yet, algorithms can also be used to influence the behavior of populations as well as individuals. A key example arises from social media, where products are generally designed to keep users engaged by using data to both predict and manage users' conduct to make more accurate and personalized predictions on future suggested content (O'Neil, 2017; Zuboff, 2019). These processes not only expand the amount of information individuals interact with, but also restricts it through its filtering of personal and personalized information (Flaxman, Goel, & Rao, 2016; Spohr, 2017). This process can potentially decrease individuals' access to and development of new ideas and new points of view as the user loses autonomy over what kind of information they are exposed to, with effects on the way people both interact and perceive one another, affecting individual identity construction (Dutton, Reisdorf, Dubois, & Blank, 2017; Levinson, Cogburn, & Vodanovich, 2018). Even by making recommendations to users, algorithms can influence behavior and preferences—and diminish the role of human decision-making—by elevating some choices or options while suppressing others, which could potentially be used for economic or political gain by state or private actors (Yeung, 2017, 2019).

These concerns are not entirely unique to neurotechnologies, as they arise from the growth of AI and digital tools used in products and services more generally. However, DTC neurotechnologies may raise particular issues around autonomy from this perspective. Algorithmic recommendations made to users based on their neural data collected by BCIs may have particular sway if the data collected speak to preferences or decision-making patterns better than other data types or available methods of data collection. Decisions made by algorithms that affect individuals or groups based on neural data—perhaps especially when autonomous decisions are made to alter brain function, as in a closed-loop BCI—create high stakes environments where algorithmic decision-making entities may and perhaps should receive more scrutiny (see Martin & Waldman, 2023).

Further, state or private actors deploying BCIs with these decision-making capabilities may craft or extend new power relationships with those subject to neurotechnology-based decisions. Consider the potential for an individual with a disability to use a wearable BCI to assist with speech. While such innovations could expand autonomy by enabling an individual to communicate information or preferences who might not otherwise be able to, such devices also raise concerns around whether the devices will provide accurate and valid expressions of the user's intents or preferences (Chandler et al., 2021; see also IBC, 2021). By making decisions about how to convert neural signals into text or audio, a BCI becomes not only an enabling device but also regulates the potential for communication and its accuracy.

Further ethical and legal concerns have been raised over whether such devices may enable various forms of harmful speech (Chandler et al., 2021, 2022). In response to such concerns, the BCI developer could prescribe—perhaps both digitally in code and legally in contractual terms of use—certain limitations on what a user can or cannot say or express, such as hate speech, threats of violence, or sexual speech towards a minor. To perform these functions, an algorithm would likely be monitoring content for perceived rule violations, enforcing these rules (possibly by disabling certain functions of the device), and potentially updating standards over time (see Tusikov, 2019; Yeung, 2018). From the perspective of algorithmic regulation, a private actor in this case would use digital means to regulate the speech and expression of a person or group of people with disabilities. Perhaps such interventions may be justified in some cases to prevent harmful speech, but nonetheless raise difficult political and policy questions about the use of technologies (and contract law, see Hilderbrand, 2022) to regulate behavior and speech rights.

Such considerations applied to wearable DTC neurotechnologies more broadly raise further questions about how such technologies, and those who develop and deploy them, may become more involved in mediating individual and group autonomy. Concerns over directly altering mental states raise notable concerns for individual autonomy and self-determination (Farahany, 2019; Ienca & Andorno, 2017; O'Shaughnessy, Johnson, Tournas, Rozell, & Rommelfanger, 2022). Yet, the examples above, presented through an algorithmic regulatory lens, illustrate that DTC neurotechnologies may enable

Algorithmic regulation 155

various forms of more subtle limitations on autonomy by using digital tools and infrastructure, which may nonetheless have profound impacts on personal and group decision-making. While concerns for more direct forms of manipulation may create calls for some types of policy responses, the more subtle forms of autonomy restrictions that an algorithmic regulation lens makes visible may call attention to issues to different facets of power and could require different forms of response options.

4.3 Personality and identity

Once individuals have encountered surveillance and infringement over decision making, additional concerns over potential changes to personality and identity itself emerge (IBC, 2021; Ienca & Andorno, 2017). These concerns in the neuroethical literature, again, often arise from the potential for neurotechnologies to directly alter neural states or signals, which does indeed pose ethical and policy issues. However, more subtle forms of personality and identity erosion may also be possible without direct, electrical stimulation.

Algorithmic tools increasingly aim to –and may require users to agree to– "personalize" experiences, promising further data collection for better analytics for the developers and the users' improved advertisement experiences (see Microsoft, 2018). Fan and Poole however, suggest this is an "intuitive but slippery" practice (Fan & Poole, 2006; 183), as it is assumed to not need explanation or justification before being deployed on the user and can be mobilized in any context. They also point out that this "personalization" often occurs without the direct guidance from the subject of these practices, rather being implemented by algorithms and firms. These systems of personalization, driven by algorithmic decision-making and recommendations, have the potential to influence and steer not only behaviors but also potentially preferences and therefore self-concept (Yeung, 2017, 2019; Zuboff, 2019). Social media as a case also illustrates how a mix of algorithmic dynamics and the natural desire for people to establish cognitive symmetry with others can lead to echo chamber formation, again affecting individual or group identity (Bobok, 2016; Dutton et al., 2017; Flaxman et al., 2016). These social media echo chambers may change how individuals interact with and perceive one another, and how they wish to be perceived, which can play a significant role in identity construction (Levinson et al., 2018; Pan et al., 2017).

Algorithmic decision-making paired with DTC neurotechnologies creates issues in this domain of identity as well. Wearable BCIs able to access and interpret neural data may have deeper access to mental states than data collected from currently available sources such as geolocation, search engine requests, or records of online purchases. Private or state actors aiming to use this access to neural data to make algorithmically mediated suggestions for products, services, content, or connections could pursue new techniques here for making recommendations more persuasive—potentially with anticompetitive or undemocratic political or economic goals. Neural data indicating an individual or group's response to certain kinds of advertisements or content could similarly and iteratively be used in algorithmic decision-making systems to improve the influence of these recommendations. With the potential for neural data to reveal information about an individual's identity (see eg. Ziogas et al., 2023), DTC neurotechnologies could enable not only new understandings of identity and the self, but also new ways of targeting populations with marginalized identities. DTC neurotechnologies could in some cases enable new forms of regulating populations by surveilling for marginalized identities and enabling those sorted to be targeted by other forms of policing.

5. Conclusion

Neuroethical concerns around the uniquely sensitive type of data collected by DTC neurotechnologies warrant deep analysis and consideration. Alongside other tools for ethical and policy analysis, the concept of algorithmic regulation offers potential new insights and draws attention to more structural dynamics involved in how neurotechnologies can be used to regulate individuals and populations. These insights may offer new bridges between the important neuroethical scholarship and policymaking on these issues. Algorithmic regulation can occur not only through direct stimulation of human neurology, but also through more subtle forms of algorithmically mediated influence, with potential political economic consequences. These conceptual tools can be used to revisualize or reinterpret fundamental concerns identified by the neuroethical literature around, for example, a user's identity and autonomy, to draw out different nuances and power dynamics and provide a different perspective by which to better tailor legal or regulatory interventions. This shift towards treating algorithms as regulators or regulatory tools aligns with ongoing scholarly conversations around AI and other digital technologies, potentially enabling more exchange between discussions on AI and neurotechnologies.

References

Aneesh, A. (2009). Global labor: Algocratic modes of organization. *Sociological theory, 27*(4), 347–370.

Bellanova, R., & De Goede, M. (2022). The algorithmic regulation of security: An infrastructural perspective. *Regulation & Governance, 16*(1), 102–118.

Black, J. (2002). Critical reflections on regulation. *Australasian Journal of Legal Philosophy, 27*(2002), 1–46.

Bobok, D. (2016). Selective exposure, filter bubbles and echo chambers on Facebook. *Central European University Department of Political Science. Diakses dari.* https://doi.org/10.1108/01443580911001797.

Braithwaite, J. (2008). *Regulatory capitalism: How it works, ideas for making it work better.* Edward Elgar Publishing.

Braithwaite, J., & Drahos, P. (2000). *Global business regulation.* Cambridge University Press.

Brownsword, R. (2005). Code, control, and choice: Why East is East and West is West. *Legal Studies, 25*(1), 1–21.

Brownsword, R. (2019). *Law, technology and society: Reimagining the regulatory environment.* Routledge.

Burris, S., Drahos, P., & Shearing, C. (2005). Nodal governance. *Australasian Journal of Legal Philosophy, 30*, 30–58.

Chandler, J. A., Van der Loos, K. I., Boehnke, S., Beaudry, J. S., Buchman, D. Z., & Illes, J. (2022). Brain Computer Interfaces and Communication Disabilities: Ethical, legal, and social aspects of decoding speech from the brain. *Frontiers in Human Neuroscience, 16*, 841035.

Chandler, J. A., Van der Loos, K. I., Boehnke, S. E., Beaudry, J. S., Buchman, D. Z., & Illes, J. (2021). Building communication neurotechnology for high stakes communications. *Nature Reviews. Neuroscience, 22*(10), 587–588.

Citron, D. K., & Solove, D. J. (2021). Privacy harms. *Boston University Law Review, 102*, 793–863.

Citron, D. K. (2023). Foreword, Join the Fight for Intimate Privacy. *European Union Data Protection Law Review, forthcoming, Virginia Public Law and Legal Theory Research Paper* 2023-07.

Cohen, J. E. (2019). *Between truth and power: The legal constructions of informational capitalism.* Oxford University Press.

Dutton, W. H., Reisdorf, B., Dubois, E., & Blank, G. (2017). *Social shaping of the politics of internet search and networking: Moving beyond filter bubbles, echo chambers, and fake news.*

Eyert, F., Irgmaier, F., & Ulbricht, L. (2022). Extending the framework of algorithmic regulation. The Uber case. *Regulation & Governance, 16*(1), 23–44.

Fan, H., & Poole, M. S. (2006). What is personalization? Perspectives on the design and implementation of personalization in information systems. *Journal of Organizational Computing and Electronic Commerce, 16*(3-4), 179–202.

Farahany, N. A. (2023). *The battle for your brain: defending the right to think freely in the age of neurotechnology.* St. Martin's Press.

Farahany, N. A. (2019). The cost of changing our minds. *Emory Law Journal, 69*(1), 75–110.

Flaxman, S., Goel, S., & Rao, J. M. (2016). Filter bubbles, echo chambers, and online news consumption. *Public Opinion Quarterly, 80*(S1), 298–320.

Ford, C. (2017). *Innovation and the state: Finance, regulation, and justice.* Cambridge University Press.

Goering, S., Klein, E., Specker Sullivan, L., Wexler, A., Agüera y Arcas, Bi, B., ... Yuste, R. (2021). Recommendations for responsible development and application of neurotechnologies. *Neuroethics, 14*(3), 365–386.

Goyal, N., Howlett, M., & Taeihagh, A. (2021). Why and how does the regulation of emerging technologies occur? Explaining the adoption of the EU General Data

Protection Regulation using the multiple streams framework. *Regulation & Governance, 15*(4), 1020–1034.

Haggerty, K. D., & Ericson, R. V. (2000). The surveillant assemblage. *British Journal of Sociology, 51*(4), 605–622.

Henne, K. (2019). Surveillance in the name of governance: Aadhaar as a fix for leaking systems in India Information, technology and control in a changing world: Understanding power structures in the 21st century, 223–245.

Herrera-Ferrá, K., et al. (2023). Contextual and cultural perspectives on neurorights: Reflections toward an international consensus. *AJOB neuroscience, 14*(4), 360–368.

Hilderbrand, J. R. (2022). If a social media platform was an intersection, should contract law or free speech have the right-of-way with respect to user content? *Journal of Business & Technology Law, 18*, 171.

Ienca, M., & Andorno, R. (2017). Towards new human rights in the age of neuroscience and neurotechnology. *Life Sciences, Society and Policy, 13*(1), 1–27.

Ienca, M., Haselager, P., & Emanuel, E. J. (2018). Brain leaks and consumer neuro-technology. *Nature Biotechnology, 36*(9), 805–810.

Ienca, M., & Malgieri, G. (2022). Mental data protection and the GDPR. *Journal of Law and the Biosciences, 9*(1) art. no. lsac006.

International Bioethics Committee. (2021). *Report of the International Bioethics Committee of UNESCO (IBC) on the ethical issues of neurotechnology.* https://unesdoc.unesco.org/ark:/48223/pf0000378724.

Koop, C., & Lodge, M. (2017). What is regulation? An interdisciplinary concept analysis. *Regulation & Governance, 11*(1), 95–108.

Latour, B. (2005). *Reassembling the social: An introduction to actor-network theory.* Oxford University Press.

LeCun, Y., Bengio, Y., & Hinton, G. (2015). Deep learning. *Nature, 521*(7553), 436–444.

Lessig, L. (1999). *Code: And other laws of cyberspace.* Basic Books.

Levinson, N., Cogburn, D. L., & Vodanovich, S. (2018). *Introduction to the minitrack on social media: Culture, identity, and inclusion.*

Li, J. H., & Zhao, Y. Y. (2014). Improvement and simulation of artificial intelligence algorithm in special movements. *Applied Mechanics and Materials, 513*, 2374–2378.

Martin, K., & Waldman, A. (2023). Are algorithmic decisions legitimate? The effect of process and outcomes on perceptions of legitimacy of AI decisions. *Journal of Business Ethics, 183*(3), 653–670.

Neurosky. (2023). *Headset EEG.* https://store.neurosky.com/.

O'Neil, C. (2017). Weapons of math destruction: How big data increases inequality and threatens democracy. *Crown.*.

O'Shaughnessy, M., Johnson, W. G., Tournas, L., Rozell, C., & Rommelfanger, K. (2022). *Neuroethics guidance documents: Principles, analysis, and implementation strategies* Available at SSRN 4035992.

Pan, Z., et al. (2017). Who do you think you are? Common and differential effects of social self-identity on social media usage. *Journal of Management Information Systems, 34*(1), 71–101.

Rommelfanger, K. S., Jeong, S. J., Ema, A., Fukushi, T., Kasai, K., Ramos, K. M., & Singh, I. (2018). Neuroethics questions to guide ethical research in the international brain initiatives. *Neuron, 100*(1), 19–36.

Rommelfanger, K. S., et al. (2019). Neuroethics: think global. *Neuron, 101*(3), 363–364.

Sayes, E. (2014). Actor–network theory and methodology: Just what does it mean to say that nonhumans have agency? *Social Studies of Science, 44*(1), 134–149.

Sell, S. K. (2022). Twenty-first-century capitalism: A research agenda. *Global Perspectives, 3*(1), 35540.

Spohr, D. (2017). Fake news and ideological polarization: Filter bubbles and selective exposure on social media. *Business Information Review, 34*(3), 150–160.

Tusikov, N. (2019). Regulation through 'bricking': Private ordering through the 'Internet of Things'. *Internet Policy Review, 8*(2), https://doi.org/10.14763/2019.2.1405.

Ulbricht, L., & Yeung, K. (2022). Algorithmic regulation: A maturing concept for investigating regulation of and through algorithms. *Regulation & Governance, 16*(1), 3–22.

UNESCO. (2023). *Unveiling the neurotechnology landscape: Scientific advancements, innovations and major trends.* https://unesdoc.unesco.org/ark:/48223/pf0000386137.

Wexler, A., & Reiner, P. B. (2019). Oversight of direct-to-consumer neurotechnologies. *Science (New York, N. Y.), 363*(6424), 234–235.

Whitley, E. A., Gal, U., & Kjaergaard, A. (2014). Who do you think you are? A review of the complex interplay between information systems, identification and identity. *European Journal of Information Systems, 23*, 17–35.

Yeung, K. (2018). Algorithmic regulation: A critical interrogation. *Regulation & Governance, 12*(4), 505–523.

Yeung, K. (2017). 'Hypernudge': Big Data as a mode of regulation by design. *Information Communication & Society, 20*(1), 118–136.

Yeung, K. (2019). 'Hypernudge': Big Data as a mode of regulation by design. *The social power of algorithms.* Routledge, 118–136.

Ziogas, A., Mokros, A., Kawohl, W., de Bardeci, M., Olbrich, I., Habermeyer, B., & Olbrich, S. (2023). Deep learning in the identification of electroencephalogram sources associated with sexual orientation. *Neuropsychobiology, 82*(4), 234–245.

Zuboff, S. (2019). *The age of surveillance capitalism: The fight for a human future at the new frontier of power.* Hachette Book Group.

Further reading

AI, H. (2019). High-level expert group on artificial intelligence. *Ethics guidelines for trustworthy AI, 6.*

European Group on Ethics in Science and New Technologies. (2018). *The European Group on Ethics in Science and New Technologies: An ethical, societal and fundamental rights dimension for the EU policies.*

Floridi, L., Cowls, J., Beltrametti, M., Chatila, R., Chazerand, P., Dignum, V., & Vayena, E. (2021). An ethical framework for a good AI society: Opportunities, risks, principles, and recommendations. *1*(25), 19–39.

Jobin, A., Ienca, M., & Vayena, E. (2019). The global landscape of AI ethics guidelines. *Nature Machine Intelligence, 1*(9), 389–399.

Neto, J. A. R. (2020). Data generated by wearables. *Medium.* https://medium.com/xnewdata/data-generated-by-wearables-48da42a88263#.

Ryan, M., & Stahl, B. C. (2020). Artificial intelligence ethics guidelines for developers and users: clarifying their content and normative implications. *Journal of Information Communication and Ethics in Society, 19*(1), 61–86.

Van Est, Q. C., Gerritsen, J., & Kool, L. (2017). Human rights in the robot age: Challenges arising from the use of tics, artificial intelligence, and virtual and augmented reality.

CHAPTER NINE

The extended mind thesis and the cognitive artifacts approach: A comparison

Guido Cassinadri[a,*] and Marco Fasoli[b]

[a]Institue of Health Science, Sant'Anna School of Advanced Studies, Pisa, Italy
[b]Department of Philosophy, University of Roma La Sapienza, Rome, Italy
*Corresponding author. e-mail address: guido.cassinadri@santannapisa.it

Contents

1. Introduction	162
2. The extended cognition and the cognitive artifacts approach	163
2.1 EXT: Extended mind and extended cognition together again	163
2.2 Cognitive artifacts	165
3. How to extend cognition, cognitive artifacts, and AI	167
3.1 The mark of cognition and matrix of cognition approaches	167
3.2 Different approaches to cognitive artifacts	172
3.3 CA, EXT, and AI—Comparison, intersection and moral implications	173
4. Conclusion: Compatibility and differences between EXT and CA and some ethical implications	177
References	179

Abstract

In this chapter, we will present and compare two theoretical approaches that focus on the use of artifacts for cognitive purposes—the extended mind thesis (EXT) and the cognitive artifacts approach (CA)—and analyze these differences in order to highlight the strengths and flaws of each approach. According to the extended mind thesis in some cases the tools are so integrated into the cognitive system that they have to be considered an integral part of the mind of the subject. According to the competing cognitive artifacts approach, artifacts that play a crucial role in cognition have to be identified as a specific class of artifacts called "cognitive artifacts". Having introduced both approaches, we will then respectively present the main argumentative strategies that have been used to support EXT, and the main taxonomies of cognitive artifacts that have been proposed in the literature. To conclude, we will analyze the relations between EXT and CA as well as their ethical implications, showing that although they are not mutually exclusive, they offer different perspectives on very similar phenomena.

Developments in Neuroethics and Bioethics, Volume 7
ISSN 2589-2959, https://doi.org/10.1016/bs.dnb.2024.02.004
Copyright © 2024 Elsevier Inc. All rights reserved, including those for text and data mining, AI training, and similar technologies.

Abbreviations

EXT   Theory of Extended Mind and Extended Cognition
EMB   Theory of Embedded Mind and Embedded Cognition
CA   Cognitive Artifacts Approach
S1   System 1
S2   System 2

1. Introduction

Until the 1990s, cognitive sciences focused almost entirely on what happened inside the brain and what potentially could be reproduced by a computer, considering cognition as something that it was possible to study separately from its environment. Classical cognitive science was committed to an internalist (brain-bounded) view of cognition, conceived as a representational symbol processing device that receives inputs from the sensory organs and delivers behavioral output to the body. While classical cognitive science assumed a disembodied, abstract, detached and internalist version of cognition, the 4E approaches shifted the paradigm towards a view of cognition as *embodied, embedded, enactive,* and *extended* (Newen, De Bruin, & Gallagher, 2018). The common denominator of 4E can be identified in the rejection of the classical Sandwich Model of Cognition[1] (Gallagher, 2023; Hurley, 1998). According to this metaphor, classical cognitive science posited that information processing involved three distinct and independent domains: high-level cognition, perception, and action. While high-level cognition represented the meat, perception and action represented the two pieces of low-protein bread that traditional cognitive scientists discarded to concentrate solely on the first ingredient, considered a specific domain of the brain. Bodies, actions, the environment, and their relationship with traditional aspects of cognition were instead the favored subjects of investigation by 4E cognitive scientists. In this chapter we compare two approaches within the family of 4E cognition, both focusing on a specific part of the environment, namely technological artifacts, by examining their role in cognition and their ethical implications. These approaches are the extended mind model (EXT) presented in Section 2.1, and the cognitive artifacts approach (hereafter CA) presented in Section 2.2. Both approaches can be applied to AI-based technological artifacts, providing a theoretical framework that could be useful for addressing their ethical aspects.

---

[1] Nevertheless, there are some exceptions among the proponents of embodied cognition, who have not completely abandoned the sandwich model (Shapiro, 2014).

According to EXT, under specific circumstances some objects play such an essential role in cognition that they constitute part of the agent's cognitive system, even if they are not physically part of the brain. According to this perspective, these objects are not merely important for human cognitive processes; they are an integral part of the cognitive machinery that extends beyond confines of the skull. In Section 3.1 we present the main argumentative strategies used to support this assertion.

As already noted, the extended mind is not the only approach that gained prominence in the 1990s and emphasized the significance of external objects in cognitive processes (Norman, 1993). Another, albeit smaller-scale debate is based on the idea that objects particularly relevant for cognitive processes should be regarded as ontologically distinct from traditional artifacts, warranting a new and different label: "cognitive artifacts". Cognitive artifacts have been defined as "physical objects made by humans for the purpose of aiding, enhancing, or improving cognition" (Hutchins, 1999, p. 126), and in Section 3.2 we present the main taxonomies of this class of objects.

Although they focus on similar phenomena, EXT and CA have received very different levels of attention from scholars. CA has often been associated with the distributed model of cognition and has rarely been discussed by proponents of EXT. What are the principal theoretical differences and similarities between the two approaches? In this chapter we will address this fundamental question.

2. The extended cognition and the cognitive artifacts approach

2.1 EXT: Extended mind and extended cognition together again

The originality of EXT consists in offering an original answer to the question "Where does the mind end and the rest of the world begin?" (Clark & Chalmers, 1998). This question arises from the view that if we embrace a functionalist view of the mind, defining the ontology of mental states and cognitive processes based on the functional role they play within a system (Putnam, 1975), then there is no inherent reason to confine them to the enclosed space of our brains. It is plausible that the artifacts we utilize, such as smartphones and notebooks, can, under certain circumstances, precisely fulfill the functional role traditionally ascribed to neural resources.

Clark and Chalmers (1998) introduced the extended mind thesis through the well-known thought experiment of Otto and Inga. Inga, a neurotypical human cognitive agent, possesses direct access to her dispositional beliefs stored in her biological memory. Consequently, when she wishes to visit the Museum of Modern Art in NY, she accesses the information about its address stored in her brain and behaves accordingly. Otto on the other hand suffers from a specific form of Alzheimer's disease and relies on a notebook to store essential information for his daily life. When he learns new relevant information, he notes it down in his notebook and retrieves the information when needed, subsequently acting accordingly. The authors argued that, despite the differing ways of accessing and utilizing stored information, these cases are entirely analogous in all relevant functional respects. For Otto, his notebook serves the same coarse-grained functional role that biological memory has for Inga, driving their behavior on both an explanatory and ontological level of analysis (Chalmers, 2008; Clark, 2008). Thus, the notebook's encodings function as an external and extended material basis of Otto's memory and as a physical vehicle for his memory processes and dispositional beliefs. This conclusion was originally based on the parity principle:

If, as we approach a task, a part of the world functions as a process which, were it done in the head, we would have no hesitation in recognizing as part of the cognitive process, then that part of the world is (so we claim) part of the cognitive process. (Clark & Chalmers, 1998, p. 8).

Thus, EXT calls into questions not only the location of the mind, traditionally understood to be confined to the head, but also the initially presumed internal material substrate of mental states and cognitive processes. If what matters is their functional role (Putnam, 1975), then these processes need not be exclusively realized by biological matter but can also be realized through the interconnected interplay between the brain, bodily actions and external devices (Wheeler, 2011).

It is important to clarify that Clark and Chalmers (1998) originally proposed the extended mind thesis and the extended cognition thesis separately. While the former concerns folk psychology types, such as states of dispositional beliefs, desires, and preferences, and considers their role in guiding behavior, the latter applies to external supports that functionally contribute to cognitive processes (Wheeler, 2019). A paradigmatic case of an extended cognitive process is one in which an arithmetic calculation is performed using pen and paper. A feedback-loop process encompassing brain operations, sensorimotor interactions with the external resources and the symbols written on the page gives rise to an extended cognitive process. However, by using the label "EXT" we can encompass both theses, referring to Chalmers' recent reformulation of EXT:

A subject's cognitive processes and mental states can be partly constituted by entities that are external to the subject, in virtue of the subject's interacting with these entities via perception and action. (Chalmers, 2019, p. 13).

This formulation elucidates what is distinctly original, disruptive and controversial about EXT, namely, its overcoming of the so-called "cognitive sandwich" of perception-cognition-action (Hurley, 1998) by incorporating specific forms of sensorimotor interactions into the cognitive domain. However, it does not capture devices like BCIs (brain-computer interfaces) that are directly integrated into the brain, mainly for therapeutic purposes (Gilbert, Ienca, & Cook, 2023), and that can arguably extend cognition (Zilio, 2020; Heinrichs, 2021). This is why we can distinguish two versions of EXT: the action-perception loops and the extended circuit thesis (Chalmers, 2019; Farkas, 2012). The first is the one realized by Otto and his notebook and it involves a sensory-motor interaction with an entity located outside the organism of the agent. The second applies instead to devices that are directly integrated to the brain processes of the subject, thus contributing to the realization of cognitive processes. This latter kind of cognitive extension is less controversial because it does not challenge the "cognitive sandwich" and is accepted as a plausible form of EXT even by major opponents of the action-perception loop thesis (Adams & Aizawa, 2008; Rupert, 2009). Moreover, this version of EXT may arguably apply for certain types of AI-based BCIs, increasingly employed in experimental and clinical settings to predict and alleviate symptoms of various neurological and psychiatric disorders (Gupta, Vardalakis, & Wagner, 2023).

2.2 Cognitive artifacts

Technological artifacts have often been conceived as ways to enhance human abilities, which in many respects are limited. Vehicles make us faster, glasses and microscopes improve our vision, mass media allows us to communicate faster, and so on. As anticipated in the introduction, the notion of a cognitive artifact has been elaborated to identify a subset of artifacts that play a crucial role in cognition. In addition to Hutchins' definition provided above, also Heersmink provides a useful perspective: cognitive artifacts are "objects that contribute functionally to the performance of a cognitive task" (Heersmink, 2013, p. 46). Exemplary instances of cognitive artifacts include shopping lists, abacuses, GPS navigators, maps, and more.

Basically, under specific circumstances, cognitive artifacts play a pivotal role in cognition by providing users with information via diverse representations. In turn, these representations enable users to exploit different cognitive mechanics beneficial for solving a problem. According to Casati (2017), some types of cognitive mechanics exploited by cognitive artifacts are: "freeing up working memory, storing information, making inferences visually available and available for inspection, priming action, bridging different modules, storing information." (Casati, 2017, p. 14). For example, an abacus frees up working memory necessary for solving a mathematical problem, physicians use checklists to store information and avoid oversights, diagrams make some inferences visually available, and maps assist users in identifying the shortest route between two points. Heersmink (2013) notes that some cognitive artifacts function by providing users information in a non-representational way, whereas others (Fasoli, 2018a; Miłkowski, 2022) have argued that these cases can also be considered representational, because they ultimately play a representational role.

Different cognitive artifacts may push users towards different cognitive processes when tackling in the same task. Heersmink (2021), for instance, discusses the empirical results of an experiment (Mueller & Oppenheimer, 2014) comparing note-taking processes with a laptop versus traditional pen and paper. The data suggests that, since typing is quicker than writing with a pen, a laptop encourages users to take down the teacher's lesson verbatim, allowing for more information storage by using an artifact. However, the study also compared the performances of students and found that, counterintuitively, forcing users to selectively choose relevant information for storage using pen and paper may be more beneficial for their memorization process. In other words, the results indicate a "less is more" scenario, where being capable of storing more information may actually be counterproductive.

Cognitive artifacts have been the object of study of both cognitive scientists (Norman and Hutchins) and philosophers (Heersmink, Casati and Fasoli). The former focused primarily on how cognition may be distributed through cognitive artifacts and how their design affects the user and the tasks they perform. The latter focused on theoretical aspects such as the ontological status of cognitive artifacts, taxonomies to classify them and ethical problems arising from their use. Regarding ethical concerns, several of them are tied to the so-called cognitive diminishment hypothesis, suggesting the possibility of diminished human capacities caused by overreliance on these objects or by other phenomena, as illustrated in the case of note-taking with a laptop.

From an ontological point of view, we can note that the debate concerning cognitive artifacts has not been explicitly linked to the debate on realism or eliminativism concerning artifacts (Preston, 2013; Thomasson, 2007). Rather, from the very beginning the notion of a cognitive artifact implicitly assumes a realistic stance towards artifacts. Defending the ontological status of cognitive artifacts while simultaneously adopting an anti-realistic position towards other kinds of artifacts would likely be peculiar and probably self-contradictory. The philosophical debate concerning cognitive artifacts has instead focused on the notion of function (Fasoli, 2018b; Heersmink, 2014), and has distinguished several kinds of functions that these objects can fulfill. A prominent distinction often discussed is between artifacts with selected functions and artifacts with improvised functions (Heersmink, 2014). Sometimes we use cognitive artifacts that have been intentionally designed for cognitive purposes, while in other circumstances we attribute cognitive function in a creative manner, for instance when we take notes on a napkin.

In conclusion, it is intriguing to observe that the relationship of the CA with the 4E cognition frameworks seems somewhat controversial, at least according to conventional conceptions of 4E cognition approaches. On the one hand, the CA focuses on the role of objects that are outside the brain, in the world (exactly as EXT), for cognition, entirely consistent with 4E frameworks in this respect, and specifically with the distributed cognition approach (Hutchins, 1995). On the other hand, postulating cognitive artifacts implies a sharp separation of some specific objects characterized as "cognitive", implicitly assuming a notion of cognition as high-level cognition (Fasoli, 2018b), from others labeled as "non-cognitive artifacts". However, among the non-cognitive artifacts, one may likely find several artifacts that are discussed and analyzed by other 4E cognitive scientists, particularly by those more oriented towards the paradigm of embodied cognition (Schettler, Raja, & Anderson, 2019). For this reason, it appears that the CA is still committed to the cognitive sandwich.

3. How to extend cognition, cognitive artifacts, and AI
3.1 The mark of cognition and matrix of cognition approaches

In this section we will consider the mark of cognition and the matrix of cognition as two argumentative strategies that have been used to support EXT in contrast to the Embedded Mind thesis (EMB), which posits that the mind is merely embedded and causally supported by external resources (Rupert, 2004, 2009). As we explore these strategies, we will also describe the shift from the first to the second wave of EXT, as identified by John Sutton (2010), as two distinct but historically overlapping strategies for arguing in favor of EXT. It is challenging to draw a clear distinction between the two waves by using Sutton (2010) criterion—namely the shift from the parity principle to the complementarity principle—since Clark himself has always been committed to the latter (Clark, 1997, p. 68, 1998, p. 99). Therefore, we identify the first wave only with Clark and Chalmers (1998) seminal paper, showing the evolution of argumentative strategies supporting EXT into the second wave, reaching its apex in Heersmink (2015) framework.

Critics early faulted EXT for not offering a precise definition of the mind or cognition. Rather than providing explicit definitions, Clark and Chalmers appealed to an implicit understanding of these notions, later defined as what enables "intelligent behavior" (Clark, 2010a, p. 54; Clark, 2010b, pp. 92–93). Therefore, critics such as Adams and Aizawa (2001, 2008) argued that, before using the parity principle and deciding whether cognition is extended or not, a clear definition of cognition must be established. In other words, we must offer a mark of cognition: a definition of the necessary and/or sufficient conditions for an item to be part of a mental-cognitive process and system (Rupert, 2004, 2009; Walter, 2010). The need for a mark of cognition to support EXT has been emphasized by both critics (Adams & Aizawa, 2008, 2010) and proponents (Kersten, 2022; Piredda & Di Francesco, 2020; Rowlands, 2009, 2010; Wheeler, 2010, 2011, 2016, 2019) of EXT. The latter aimed to devise marks of cognition capable of encompassing both internal and external resources as constitutive components of cognitive processes and systems. However, in line with other critics (Allen, 2017; Facchin, 2023; Varga, 2017), Clark has always expressed sceptisicm about the possibility of identifying marks of cognition, given the extensive heterogeneity of its forms (Clark, 2011, p. 452).

The first wave of EXT relied on three elements: an intuitive grasp of cognition (Clark, 2011, 2019), the parity principle, and the coupling conditions. While interpretations of the parity principle may vary (Cassinadri, 2022, p. 5; Wheeler, 2011), its cornerstone lies in the notion that if X functionally operates, at a specific level of analysis, akin to what is recognized as paradigmatically cognitive—namely cognitive-enabling brain processes—, then X can be part of an extended cognitive system (Facchin, 2022). While Clark (1998, 2008, 2019) consistently considered a coarse-grained level of analysis,

The extended mind thesis and the cognitive artifacts approach

his critics stressed that on a fine-grained level, this isomorphism is difficult to find (Adams & Aizawa, 2008; Rupert, 2004; Adams & Aizawa, 2010). To settle the issue, Wheeler (2011), pp. 419–420 argued that parity should not be conceived "as parity with the inner simpliciter, but rather as parity with the inner with respect to a scientifically informed, theory-loaded, locationally uncommitted account of the cognitive", namely a mark of the cognitive.

Along with the parity principle, Clark and Chalmers introduced "glue and trust" coupling conditions between the agent and the external resource to prevent so-called "cognitive-bloat", namely the excessive expansion of cognition into the environment. According to these conditions, an informational resource can become part of the agent's extended mind if and only if there is a reliable coupling between the agent and the resource, ensuring that the resource is (1) constantly used in the agent's life, (2) readily available without difficulty, (3) automatically trusted, and (4) endorsed consciously in the past (Clark & Chalmers, 1998; Clark, 2008). Responses to these coupling conditions can take two forms. The first approach views phenomenological coupling conditions as unnecessary for cognitive extension (Facchin, 2022; Smart, Andrada, & Clowes, 2022). Alternatively, functional and phenomenological coupling conditions can be understood as potentially open-ended dimensions, with parameters set on a continuum rather than discrete criteria. This is the direction taken by what we call "the matrix of cognition" approach, which posits that cognitive systems are characterized by an open-ended set of cognitive dimensions that can be set on different parameters. These dimensions characterize the internal resources and their relations (*cognitive features*) as well as the relations between internal and external resources. Unlike in a finite set of coupling conditions, these dimensions are an open-ended set of *dimensions of integration*.

The path toward the second wave of EXT therefore originated from another criticism affecting the first wave, namely the nature and potential arbitrariness of the glue and trust conditions. Why exactly should constancy, availability, and reliability of the external informational resource be exclusively identified as the relevant agent-artifact coupling conditions that make an external tool cognitive? If an agent does not immediately trust an item of his internally stored information, should this count as non-cognitive? Clark (2010b), p. 83 always noted that coupling conditions do not discretely make something cognitive per se but rather gradually make an object a proper part of a cognitive system. Moreover, the key point of the matrix of cognition approach is that no *cognitive feature* per se characterizes the internal resources and their relations, and no *coupling condition* per se should be elevated as a necessary and/or sufficient mark of cognition and cognitive extension.

Within the matrix of cognition approach, both cognitive features, namely the properties of the internal resources, as well as the dimensions of integration between the agent and the external resource can be conceived as "cognitive dimensions" (Heersmink, 2015, p. 583). Applying this matrix of cognition approach, Clark (2010a), p. 62 argues that even if the memory processes of a Martian were slower than ours, setting their cognitive feature of "timing" in remembering on a different parameter along a continuum, this difference would not disqualify their form of remembering from being genuinely cognitive. Similarly, a subject coupled to a notebook or neural implant might exhibit an artificially extended form of remembering with different temporal sequences and no recency or primacy effects (Rupert, 2004; Weiskopf, 2008). Clark (2005, 2010a) used this kind of argument based on the matrix of cognition approach to respond to his critics (Gertler, 2007; Rupert, 2004; Sterelny, 2004)[2] and compensate for the lack of a scientifically informed mark of cognition.

Starting from the mid-2000s, some authors developed the EXT-second wave in two stages. Since the glue and trust conditions did not exhaust all the dimensions of integration between the agent and the tool, these authors identified the relevant functional and phenomenological dimension of cognitive integration between the agent and the tool. This shift abandoned the parity principle in favor of the complementarity principle (Heersmink, 2012, 2015; Menary, 2006, 2007, 2010a; Sutton, 2006, 2010):

In extended cognitive systems, external states and processes need not mimic or replicate the formats, dynamics, or functions of inner states and processes. Rather, different components of the overall (enduring or temporary) system can play quite different roles and have different properties while coupling in collective and complementary contributions to flexible thinking and acting. (Sutton, 2010, p. 194).

Crucial elements of this shift were contributed by Sutton, Sterelny and Heersmink. Sutton (2006, 2010) provided a taxonomy of the relevant cognitive dimensions of integrations between the agent and the tool. Sterelny (2010) added the dimensions of trust, individualization, and entrenchment, that is the reciprocal transformation of internal and external

---

[2] For example, Clark (2010a, p. 57) used the same argument about the dimension of "reliability" of information in memory retrieval to reply to Sterelny (2004) and Gertler (2007).

EXT-first wave (Clark and Chalmers 1998)	EXT- second wave (Heersmink 2015)
Parity Principle: Identifies a functional isomorphism between internal and external resources.	**Complementarity Principle**: Highlights the differences between internal and external resources.
Glue and Trust Coupling Conditions: Isolates specific dimensions of integration of the matrix of cognition between the agent and the tool.	**Agent-tool Dimensions of Integrations**: Open-ended set of dimensions of integration of the matrix of cognition between the agent and the tool.

Fig. 1 A comparison between the first and the second wave of EXT.

components. Heersmink (2012, 2015) reached the apex of the second wave by proposing the most encompassing open-ended set of dimensions: reliability, durability, trust, procedural and representational transparency, individualization, bandwidth, speed of information flow, distribution of computation, and cognitive and artifactual transformation. Heersmink's model envisions different degrees of cognitive integration across these dimensions, positing more or less integrated cognitive systems into different regions of a multidimensional space, thus abandoning the clear-cut distinction between embedded and extended cognitive systems (Fig. 1).

However, the dimensions of integration generally discussed in the literature (Clark & Wilson, 2009; Sterelny, 2010) and developed by Heersmink (2012, 2015) were based on portable representational devices, accessed via action and perception, and excluded devices directly implanted in the nervous system. Heinrichs (2021) offered a broadened interpretation of some dimensions, applying them to both versions of EXT. According to this position, closed-loop BCIs do not provide the users with representations[3] and can reach a high level of cognitive integration on several dimensions, such as individualization and speed of information flow. This also thanks to their AI algorithms for brain features extraction and classification, enabling a fine-grained coordination between the brain and the device, which can stimulate the former for various therapeutic purposes (Gupta et al., 2023; Valeriani, Santoro, & Ienca, 2022). Therapeutic BCIs can functionally contribute to the

---

[3] However, open-loop advisory BCI systems, like the ones produced by the company Neurovista, predict epileptic seizures and notify the patients via an external portable representational artifact (Gilbert et al., 2023). Moreover, it is worth noticing that some BCIs are coupled to digital screens to enable patients to communicate via written text, which is perceptually accessed by the user and may be interpreted in terms of EXT (Kyselo, 2013; Zilio, 2020). However, in this chapter we focus on closed-loop BCIs that deliver their output directly on the brain of patients, without involving action-perception loops.

cognitive system of patients by directly modifying their brain in such a way that it can perform the same coarse-grained function of a healthy brain (Heinrichs, 2017) . Moreover, if we consider closed-loop BCI systems, in which the control is delegated to the AI-powered device, the patients can acquire new skills and capacities otherwise impossible (Coin & Dubljević, 2021; Gilbert et al., 2023; Zilio, 2020).

3.2 Different approaches to cognitive artifacts

One of the fundamental questions that surrounds cognitive artifacts is: "how does the human mind use these artifacts in order to improve its performance?". This poses a difficult challenge due to the diverse range of objects falling under the label "cognitive artifact", each potentially employed in many different ways. One way to try and tackle this difficulty is to delve into a detailed description of these objects, such as providing a taxonomy that allows us to recognize similarities among them. At least three different theoretical approaches have been developed in order to analyze cognitive artifacts.

The first type of theoretical approach was developed by Brey (2005) and centered on the cognitive capacities of tool users. This approach seeks to classify artifacts by focusing on the cognitive capacities they facilitate (see also Fasoli, 2022). An abacus supports mathematical capabilities and can be characterized as a mathematical device; a checklist aids memory and can be characterized as a memory device, and so forth. Even though Brey (2005) talks about the "extension" of cognitive capacities through cognitive artifacts, he was not explicitly committed to EXT. His classificatory approach seems quite intuitive. Each artifact is classified with respect to the cognitive capacity it supports, and this appears to be in line with the common view of technological artifacts as aids.

A second approach as suggested by Heersmink (2013), p. 468 criticizes Brey (2005) cognitive-centered framework for its arbitrary anthropocentrism and its neglect of intrinsic artifact features, defining them only with reference to the cognitive capacities they enhance. Heersmink introduces a distinction between cognitive artifacts and cognitive techniques, the latter being procedures used for solving some cognitive tasks (e.g., mnemonics can be used as alternatives to checklists). Then, as we anticipated in Section 2.2, he distinguishes between representational and non-representational artifacts. Drawing on Peirce's work on representations, Heersmink (2013) adopts an approach that differentiates from the one elaborated by Brey insofar as it attempts to analyze cognitive artifacts independently from human capacities by focusing on the kind of representations they offer. For example, a map is an iconic representational device that can be conceived as an iconic artifact; a compass is an indical artifact, etc.

A third classificatory approach was developed by Fasoli (2018a), who tries to isolate three fundamental kinds of relationships between human cognition and the cognitive artifacts: complementarity, substitution and innovation. Some artifacts complement human capabilities (e.g. an abacus), others substitute some cognitive processes (e.g. a GPS navigation system replacing a user's sense of orientation), and yet others create cognitive tasks that would not exist without that artifact (e.g. the Tetris game). This approach can therefore be considered interaction-based. A related taxonomy was developed by Casati (2017), who bases his classificatory approach on the distinction between System 1 (S1) and System 2 (S2) as described by Kahneman (2011). S1 and S2 are two fundamental modes of functioning that characterize human cognition. S1 is automatic, fast, involuntary and rough, while S2 is slower and requires effort to be activated. For instance, recognizing someone else's emotion is a rapid process that occurs in S1, while dividing 1240 by 32 requires S2. Casati identifies two additional modes of cognition, S3 and S4, (which he calls M3 and M4) that can be considered analogous to S1 and S2 but involve the use of artifacts. S4 corresponds to S1 but with the involvement of an artifact (e.g. distracted use of a navigation tool such as Google Maps), while S3 corresponds to S2 with artifact involvement (e.g. using a paper map that needs orientation and interpretation).

Finally, we need to mention an original approach recently developed by Miłkowski (2022), who interprets scientific knowledge representations as cognitive artifacts. This is one of the few cases where the term cognitive artifact is used to refer specifically to abstract knowledge representations, which always require physical support.

Despite their differences, all these theories share a common challenge: the absence of a clear and explicit definition of cognition to underpin the notion of cognitive artifact. They rely on references to prototypical cases of high-level cognition (Fasoli, 2018b) that involve the use of artifacts, without providing a precise definition.

3.3 CA, EXT, and AI—Comparison, intersection and moral implications

We will begin our comparison with some high-level preliminary remarks. First, we aim to clarify the differences between the two versions of EXT and the CA. Since most BCIs do not provide users with representations, they can hardly be considered cognitive artifacts, making them more apt for interpretation in terms of "extended circuit" EXT. Consequently, our comparative focus will be on the "action–perception loops" of EXT and CA, which exhibit both similarities and notable differences. For instance, at first glance Otto's notebook looks like a paradigmatic example of a cognitive artifact, even if it is rarely cited as such an object. This is probably because in the EXT perspective the notebook somehow disappears as an artifact, undergoing an ontological transformation and becoming a part of the mind. This allows us to identify an important ontological difference between EXT and CA. While, according to "action–perception loop EXT", certain artifacts highly integrated with the human mind via action and perception and exhibiting specific features, become part of an extended cognitive system, CA does not imply such strong ontological stance. Instead, CA remains committed to the cognitive sandwich, distinguishing perceptual access to the device, its manipulation and internal cognitive processes (Hurley, 1998). Therefore, the ontological commitment implied by CA is more modest than that of EXT, as it simply entails a realistic stance on artifacts and the proposal to introduce a sub-class of these objects with specific cognitive properties. CA can be considered a form of conceptual engineering (Fasoli, 2022, p. 49), while EXT is first and foremost an ontological thesis (Heinrichs, 2017) .

In addition, as emphasized by Milkowski, "according to the extended mind thesis, the individual mind remains at the center of a cognitive system; this individual-centered perspective is not endorsed by proponents of distribution" (Miłkowski, 2022, p. 219). In particular, EXT-first wave appears much more aligned with a Cartesian perspective (Floridi, 2014, p. 95), since first-person perspective characterizes mentality and defines the coupling conditions with the tool (Piredda & Di Francesco, 2020).[4] However, this does not apply to the extended circuit version of EXT, wherein the brain is subpersonally accessed and manipulated[5] by a closed-loop BCI system without the user's oversight. Moreover, this is also not the case in other paradigms belonging to the 4E domain, such as cognitive

---

[4] However, some proponents of EXT stressed that the phenomenological coupling conditions do not play a relevant or necessary role for supporting the claim that cognition can extend (Facchin, 2022; Smart et al., 2022).

[5] The term should be interpreted as neutral here, meaning that the device causally influences the brain functionings.

distribution theory (Hutchins, 1995), which are even compatible with AC. While EXT focuses on the conditions of extension between the individual agent and the tool, what characterizes humans' complex capacities is that cognitive artifacts are often used as "cognitive commons", namely as shared resources that are publicly accessible, usable and transformable by a community (Sterelny, 2010). In a cognitive ecology encompassing multiple human agents, artifacts, and cultural scaffoldings, heterogenous systems interact on different levels in interdependent ways. Since "every theory implies a set of ontological commitments and every ontological commitment emphasizes some kinds of connections over others" (Hutchins, 2010, p. 706), EXT may be ill-suited to account for the interaction of multiple external resources, if it establishes rigid ontological and explanatory boundaries that constrain the explanation of complex phenomena. However, the matrix of cognition account enables us to posit more or less integrated cognitive systems involved in cognitive ecologies. Morevoer, EXT may offer a suitable framework for interpreting those BCIs that are highly integrated on relevant dimensions and functionally contribute to agents' cognitive system (Zilio, 2020; Heinrichs, 2021).

The latest trend within the EXT debate interrogates whether and how AI systems may qualify as cognitive extensions (Hernández-Orallo & Vold, 2019; Nyholm, 2023) and its potential implications (Andrada, Clowes, & Smart, 2023; Biber & Capasso, 2022; Bublitz, 2022; Cassinadri, 2024; Vold & Hernández-Orallo, 2021). EXT seems a plausible explanation for closed-loop BCI systems in which control is delegated to the AI-powered device, allowing patients to acquire new skills and capacities that would be impossible without it. In these cases, the degree of cognitive integration seems to reach a higher level compared to devices accessed via action-perception loops.

Conversely, when considering portable representational devices implementing AI systems, they can be interpreted in both EXT and CA terms. In the latter case, we may regard them as superartifacts (Fasoli, 2018b), given their computational power and intrinsic multi-functionality. Indeed, AI systems offer users a variety of new functionalities and a multitude of representations, with consequences relevant both for the cognitive performances of users and for their emotional life (Krueger & Osler, 2022).

Let us now consider the moral implications of EXT and CA and their differences. On an ontological level of analysis, CA may be in principle compatible with both EXT and EMB, depending on whether the conditions for extension are met and whether we accept that they actually extend cognition. However, CA does not imply EXT, as it assumes an embedded view of cognition (EMB) as its default framework. Thus, to analyze the moral differences between EXT and CA we can initially focus on the distinct moral implications of EXT and EMB. One of the first authors to analyze these ethical implications was Levy (2007a, 2007b, 2011). He formulated the weak and the strong version of the ethical parity principle (EPP), with the former assuming EXT and the latter EMB.

EPP (strong): Since the mind extends into the external environment, alterations of external props used for thinking are (ceteris paribus) ethically on a par with alterations of the brain.

EPP (weak): Alterations of external props are (ceteris paribus) ethically on a par with alterations of the brain, to the precise extent to which our reasons for finding alterations of the brain problematic are transferable to alterations of the environment in which it is embedded. (Levy, 2007b, p. 61).

Vold (2018) and Drayson and Clark (Draft), and Farina and Lavazza (2022) argue that the weak EPP does not provide sufficient protection to the external parts of agents' minds. Considering mind-extending tools merely as external cognitive artifacts might expose them to lesser moral consideration and attentive treatments. Thus, Vold (2018) and Farina and Lavazza (2022) assert that there is a strong moral reason to embrace EXT instead of EMB (or AC) to better protect cognitively integrated agents from manipulation and violation of their cognitive artifacts. They appeal to the so-called Extended Cognition Moral Narrative (EXT-MN) (Cassinadri & Fasoli, 2023; Cassinadri, 2022), which is an argumentative strategy of using moral arguments in favor of EXT.

Nevertheless, proponents of EXT such as Levy (2007b, 2011), p. 68 and King (2016), pp. 47–50, acknowledge that it is sufficient to use the weak EPP to guarantee adequate and sufficient protection. They argue that there is no practical difference between EMB and EXT concerning the application of the EPP. DeMarco & Ford (2014) suggest that an embedded view of cognition, implied by CA, is sufficient for considering the morally relevant reasons for which interference with a cognitive artifact is morally illegitimate. based on the functional effects that this manipulation produces on the cognitive profile of the organism-bound agent. This means that a cognitive artifact can acquire moral status despite not being a constitutive part of the agent's mind (Søraker, 2007).

In contrast, Heersmink (2017), Peters (2022), and Farina and Lavazza (2022) assume that the moral status of cognitive artifacts is contingent on their cognitive status, but consider it as a matter of degree. According to Heersmink (2017), p. 444, the more the agent is dependent and integrated irreplaceably with a cognitive artifact, the stronger the moral reasons to combat interference against such a tool. However, Cassinadri (2022) argues that there might be no clear additional moral advantage in positing extended cognitive systems, given that there is no clear threshold for EXT and for the full moral status of such artifacts and that different degrees of their moral status can be pragmatically reduced to the strength of moral reasons and obligations that the coupling gives rise to (De Grazia, 2008; Sachs, 2011).

To conclude, appealing to moral reasons to embrace EXT rather than EMB and CA does not seem a good move. The preferability seems to depend not on the inherent moral consequences of these models but on further moral assumptions (Cassinadri & Fasoli, 2023; Cassinadri, 2022).

4. Conclusion: Compatibility and differences between EXT and CA and some ethical implications

What about the compatibility or incompatibility of EXT and CA? First, we noted that there are two versions of EXT: the "action-perception loops EXT", which overcomes the "cognitive sandwich", namely the distinction between perception, cognition, and action, and the "extended circuit EXT", which concerns non-representational devices directly implanted into the brain. In Section 2.2 we showed that CA concerns representational devices and is an approach still committed to the cognitive sandwich distinction, by focusing on high level cognition, leaving out perception and embodied interaction from the domain of cognition. Thus, the "action-perception loop" EXT and CA are prima facie incompatible for their different commitments toward the cognitive sandwich model. Nevertheless, it is possible in principle to recognize the status of cognitive artifacts for some objects while at the same time arguing that in some cases they can become a part of the mind, even if they are outside of the skull.

However, as we pointed out earlier, EXT and CA are two perspectives that are oriented towards two different sides of the problem. The first is focused on identification of the boundaries of the mind, while the second is concerned with the attribution of special status to some objects and to the understanding of how they work. "Where" versus "how", hence, are the adverbs that characterize the two perspectives, and assuming one or the other means being oriented towards quite different questions and issues. This is probably why the two approaches have never communicated very much, despite being compatible. They focus on very similar phenomena, and both endorse a vision of the human being as a cyborg (Clark, 2003), namely an ecological and opportunistic agent that exploits and hybridizes with external resources in order to complete its cognitive operations. Despite these commonalities, it appears that assuming EXT or CA pushes scholars toward different questions concerning very similar phenomena. This aspect should be considered when evaluating which framework is useful for analyzing the ethical aspects of AI-based technologies.

To conclude, despite the fact that the CA and EXT debates lived their lives independently of one another, we think that the EXT literature may benefit from conceptual clarifications developed within the CA debate.[6] For example, EXT tends to blur the differences between substantive, complementary and innovative tool-use (Fasoli, 2018a) because once an extended cognitive system is in play, it doesn't matter whether the organism-bound agent is still capable of performing a cognitive task independently of the tool's contribution, as the unity of analysis of the performing system has shifted to the extended agent+tool system.

According to Clark, these distinctions may be irrelevant for specific explanatory purposes within the domain of hybrid cognitive science, which is concerned with how extended problem-solving wholes exploit internal and external resources to complete a task (Clark, 2007, 2008, 2022). Nevertheless, we think these distinctions are still relevant on normative and ethical grounds, especially if we consider the problem of cognitive diminishment (Cassinadri & Fasoli, 2023), and that Fasoli (2018a) taxonomy offers a useful lens with a more fine-grained level of analysis to address this issue.

This chapter aimed to present an overview and a comparison of two approaches to interpreting the role of artifacts in supporting cognitive operations: EXT and CA. Moreover, we discussed the relationships between these two macro-approaches by clarifying the analogies, differences, and intersections within the constellation of frameworks internal to EXT and CA.

As we have argued, the CA literature has been developed almost in isolation, both from the analytical literature on the ontological status of artifacts (Preston, 2013) as well as from the 4E cognition debate (Newen et al., 2018), and it is thus a little narrow in scope. The literature on EXT is instead vast and heterogeneous, because the notions of mind and cognition

---

[6] For example, Zilio (2020) and Cassinadri (2024) apply Fasoli's (2018a) taxonomy of cognitive artifacts to EXT to offer a fine-grained characterization of extended cognitive systems.

are crucial to many aspects and domains of some philosophical debates. Consequently, the reconceptualisation and extension of these notions have given rise to new conceptual challenges and opportunities not only in epistemology (Carter et al., 2018) and ethics (Clowes, Smart, & Heersmink, 2023), but also in potential extensions of mind-related notions such as consciousness (Chalmers, 2019; Clark, 2009; Vold, 2015) as well as emotional and affective states and processes (Colombetti & Roberts, 2015). We think that since CA and EXT represent two potentially overlapping perspectives, that nevertheless push toward different kinds of questions and implications, addressing the ethical questions of mind-artifacts interactions and integrations is better approached from both perspectives.

References

Adams, F., & Aizawa, K. (2008). *The bounds of cognition*. Wiley-Blackwell.

Adams, F., and Aizawa, K. (2010). Defending the bounds of cognition. In Menary, R. (Ed.), 2010: The Extended Mind (pp. 43–67). Cambridge MA: MIT Press.

Allen, C. (2017). On (not) defining cognition. *Synthese, 194*(11), 4233–4249.

Andrada, G., Clowes, R. W., & Smart, P. R. (2023). Varieties of transparency: Exploring agency within AI systems. *AI & Society, 38*, 1321–1331. https://doi.org/10.1007/s00146-021-01326-6.

Biber, S. E., & Capasso, M. (2022). The right to mental integrity in the age of artificial intelligence: Cognitive human enhancement technologies. In B. Custers, & E. Fosch-Villaronga (Eds.). *Law and artificial intelligence information technology and law series*. The Hague: T.M.C. Asser Press. ⟨https://doi.org/10.1007/978-94-6265-523-2_25⟩.

Brey, P. (2005). The epistemology and ontology of human-computer interaction. *Minds & Machines, 15*, 383–398.

Bublitz, J. C. (2022). Might artificial intelligence become part of the person, and what are the key ethical and legal implications? *AI & Society*. ⟨https://doi.org/10.1007/s00146-022-01584-y⟩.

Carter, J. A., et al. (2018). *Extended epistemology*. Oxford: Oxford Academic. ⟨https://doi.org/10.1093/oso/9780198769811.001.0001⟩.

Casati, R. (2017). Two, then four modes of functioning of the mind: Towards an unification of "dual" theories of reasoning and theories of cognitive artifacts. In J. Zacks, & H. Taylor (Eds.). *Representations in mind and world. Essays inspired by Barbara Tversky* (pp. 7–23). New York: Routledge.

Cassinadri, G. (2022). Moral reasons not to posit extended cognitive systems: A reply to Farina and Lavazza. Philosophy and Technology, 35, 64. https://doi.org/10.1007/s13347-022-00560-0.

Cassinadri, G. (2024). ChatGPT and the technology-education tension: Applying contextual virtue epistemology to a cognitive artifact. *Philosophy and Technology, 37*, 14. https://doi.org/10.1007/s13347-024-00701-7.

Cassinadri, G., & Fasoli, M. (2023). Rejecting the extended cognition moral narrative: A critique of two normative arguments for extended cognition. *Synthese, 202*, 155. https://doi.org/10.1007/s11229-023-04397-8.

Chalmers, D. (2008). *Forward to Clark, A. Supersizing the mind: Embodiment, action, and cognitive extension*. Oxford: Oxford University Press, ix–xvi.

Chalmers, D. (2019). Extended cognition and extended consciousness. In M. Colombo, E. Irvine, & M. Stapleton (Eds.). *Andy Clark and his critics* (pp. 9–21). Oxford: Oxford University Press.

Clark, A. (1997). *Being there: Putting brain, body and world together again.* Cambridge MA: MIT Press.

Clark, A. (1998). Author's response. In review Symposium on Andy Clark's being there. *Metascience, 7*, 95–103.

Clark, A. (2003). *Natural-born cyborgs: Minds, technologies, and the future of human intelligence.* New York: Oxford University Press.

Clark, A. (2005). Intrinsic content, active memory and the extended mind. *Analysis, 65*(1), 1–11.

Clark, A. (2007). Curing cognitive hiccups: A defense of the extended mind. *The Journal of Philosophy, 104*(4), 163–192.

Clark, A. (2008). *Supersizing the mind: Embodiment, action, and cognitive extension.* Oxford: Oxford University Press.

Clark, A. (2009). Spreading the joy? Why the machinery of consciousness is (probably) still in the head. *Mind: A Quarterly Review of Psychology and Philosophy, 118*(472), 963–993.

Clark, A. (2010a). Memento's revenge: The extended mind, extended. In R. Menary (Ed.). *The extended mind* (pp. 43–67). Cambridge, MA: MIT Press.

Clark, A. (2010b). Coupling-constitution fallacy and the cognitive kind: A reply to Adams and Aizawa. In R. Menary (Ed.). *The extended mind* (pp. 81–101). Cambridge, MA: MIT Press.

Clark, A. (2011). Finding the mind. In Book symposium on supersizing the mind: Embodiment, action, and cognitive extension (2008). *Philosophical Studies, 152*, 447–461.

Clark, A. (2019). Author's response. In M. Colombo, E. Irvine, & M. Stapleton (Eds.). *2019: Andy Clark and his critics* (pp. 266–303). Oxford: Oxford University Press.

Clark, A. (2022). Extending the predictive mind. *Australasian Journal of Philosophy,* 1–12. https://doi.org/10.1080/00048402.2022.2122523.

Clark, A., & Chalmers, D. (1998). The extended mind. *Analysis, 58*(1), 7–19.

Clark, A., & Wilson, R. (2009). How to Situate Cognition: Letting Nature Take its Course. In M. Aydede & P. Robbins (Eds.), *2009: The Cambridge Handbook of Situated Cognition* (pp. 55–96). Cambridge: Cambridge University Press.

Clowes, R., Smart, P. R., & Heersmink, R. (2023). The ethics of the extended mind: Mental privacy, manipulation and agency. In B. Beck, O. Friedrich, & J. Heinrichs (Eds.). *Neuroprosthetics: Ethics of applied situated cognition.* Berlin: Springer.

Coin, A., & Dubljević, V. (2021). The authenticity of machine-augmented human intelligence: Therapy, enhancement, and the extended mind. *Neuroethics, 14*, 283–290. https://doi.org/10.1007/s12152-020-09453-5.

Colombetti, G., & Roberts, T. (2015). Extending the extended mind: The case for extended affectivity. *Philosophical Studies, 172*, 1243–1263. https://doi.org/10.1007/s11098-014-0347-3.

De Grazia, D. (2008). Moral status as a matter of degree? *The Southern Journal of Philosophy, 46*(2), 181–198. https://doi.org/10.1111/j.2041-6962.2008.tb00075.x.

DeMarco, P. J., & Ford, P. J. (2014). Neuroethics and the ethical parity principle. *Neuroethics, 7*(3), 317–325.

Drayson, Z., & Clark, A. (Draft). Augmentation, agency, and the spreading of the mental state. *Consciousness and Cognition.* ⟨https://sites.google.com/site/zdrayson/papers⟩.

Facchin, M. (2022). Phenomenal transparency, cognitive extension, and redictive processing. *Phenomenology and Cognitive Science.* https://doi.org/10.1007/s11097-022-09831-9.

Facchin, M. (2023). Why can't we say what cognition is (at least for the time being). *Philosophy and the Mind Sciences, 4.* https://doi.org/10.33735/phimisci.2023.9664.

Farina, M., & Lavazza, A. (2022). Incorporation, transparency, and cognitive extension. Why the distinction between embedded or extended might be more important to ethics than to metaphysics. *Philosophy and Technology, 35*, 10.

Farkas, K. (2012). Two versions of the extended mind thesis. *Philosophia, 40*, 435–447.

Fasoli, M. (2018a). Substitutive, complementary and constitutive cognitive artifacts: Developing an interaction-centered approach. *Review of Philosophy and Psychology, 9*, 671–687. https://doi.org/10.1007/s13164-017-0363-2.

Fasoli, M. (2018b). Super artifacts: Personal devices as intrinsically multifunctional, meta-representational artifacts with a highly variable structure. *Minds and Machines, 28*(3), 589–604. https://doi.org/10.1007/s11023-018-9476-3.

Fasoli, M. (2022). Cognitive artifacts between cognitive sciences and the philosophy of technology. In E. Terrone, & V. Tripodi (Eds.). *Being and value in technology.* Cham: Palgrave Macmillan. https://doi.org/10.1007/978-3-030-88793-3_3.

Floridi, L. (2014). *The fourth revolution: How the infosphere is reshaping human reality.* Oxford: Oxford University Press.

Gallagher, S. (2023). *Embodied and enactive approaches to cognition (Elements in philosophy of mind).* Cambridge: Cambridge University Press, https://doi.org/10.1017/9781009209793.

Gertler, B. (2007). Overextending the mind? In B. Gertler, & L. Shapiro (Eds.). *In Arguing about the mind* (pp. 192–206). New York: Routledge.

Gilbert, F., Ienca, M., & Cook, M. (2023). How I became myself after merging with a computer: Does human-machine symbiosis raise human rights issues? *Brain Stimulation, 16*(3), 783–789. https://doi.org/10.1016/j.brs.2023.04.016.

Gupta, A., Vardalakis, N., & Wagner, F. B. (2023). Neuroprosthetics: From sensorimotor to cognitive disorders. *Communications Biology, 6*, 14. https://doi.org/10.1038/s42003-022-04390-w.

Hernández-Orallo, J., & Vold, K. (2019). *AI extenders: The ethical and societal implications of humans cognitively extended by AI. Proceedings of the 2019 AAAI/ACM Conference on AI, Ethics, and Society (AIES '19).* New York, NY: Association for Computing Machinery, 507–513. ⟨https://doi.org/10.1145/3306618.3314238⟩.

Heersmink, R. (2012). Mind and artifact: A multidimensional matrix for exploring cognition artifact relations. In J. M. Bishop, & Y. J. Erden (Eds.). *Proceedings of the 5th AISB symposium on computing and philosophy* (pp. 54–61). Birmingham: AISB.

Heersmink, R. (2013). A taxonomy of cognitive artifacts: Function, information, and categories. *Review of Philosophy and Psychology, 4*, 465–481.

Heersmink, R. (2014). The metaphysics of cognitive artifacts. *Philosophical Explorations, 19*(1), 1–16.

Heersmink, R. (2015). Dimensions of integration in embedded and extended cognitive systems. *Phenomenology and the Cognitive Sciences, 14*, 577–598.

Heersmink, R. (2017). Distributed cognition and distributed morality: Agency, artifacts and systems. *Science and Engineering Ethics, 23*(2), 431–448.

Heersmink, R. (2021). Varieties of artifacts: Embodied, perceptual, cognitive, and affective. *Topics in Cognitive Science Science and Engineering Ethics, 23*(2), 431–448.

Heinrichs, J. H. (2017). Against strong ethical parity: situated cognition theses and transcranial brain stimulation. *Frontiers in Human Neuroscience, 11*(171), 1–13.

Heinrichs, J. H. (2021). Neuroethics, cognitive technologies and the extended mind perspective. *Neuroethics, 14*(1), 59–72.

Hurley, S. (1998). Vehicles, contents, conceptual structure and externalism. *Analysis, 58*, 1–6.

Hutchins, E. (1995). *Cognition in the wild.* Cambridge, MA: MIT Press.

Hutchins, E. (1999). Cognitive artifacts. In R. A. Wilson, & F. C. Keil (Eds.). *The MIT encyclopaedia of the cognitive sciences* (pp. 126–128). Cambridge: MIT Press.

Hutchins, E. (2010). Cognitive ecology. *Topics in Cognitive Science, 2*, 705–715. https://doi.org/10.1111/j.1756-8765.2010.01089.

Kahneman, D. (2011). *Thinking, fast and slow.* New York: Macmillan.

Kersten, L. A. (2022). New mark of the cognitive? Predictive processing and extended cognition. *Synthese, 200,* 281.

King, C. (2016). Learning disability and the extended mind. *Essays in Philosophy, 17*(2), 38–68.

Krueger, J., & Osler, L. (2022). Communing with the dead online: Chatbots, grief, and continuing bonds. *Journal of Consciousness Studies, 29*(9-10), 222–252.

Kyselo, M. (2013). Locked-in syndrome and BCI—towards an enactive approach to the self. *Neuroethics, 6,* 579–591. https://doi.org/10.1007/s12152-011-9104-x.

Levy, N. (2007a). *Neuroethics: Challenges for the 21st century.* Cambridge: Cambridge University Press.

Levy, N. (2007b). Rethinking neuroethics in the light of the extended mind thesis. *American Journal of Bioethics, 7*(9), 3–11.

Levy, N. (2011). Neuroethics and the extended mind. In J. Illes, & B. J. Sahakian (Eds.). *Oxford handbook of neuroethics* (pp. 285–294). Oxford: Oxford University Press.

Menary, R. (2006). Attacking the bounds of cognition. *In Philosophical Psychology, 19*(3), 329–344.

Menary, R. (2007). *Cognitive integration: Mind and cognition unbounded.* New York: Palgrave Macmillan.

Menary, R. (2010a). Dimensions of mind. *Phenomenology and the Cognitive Sciences, 9*(4), 561–578.

Miłkowski, M. (2022). Cognitive artifacts and their virtues in scientific practice. *Studies in Logic, Grammar and Rhetoric, 67*(1), 219–246. https://doi.org/10.2478/slgr-2022-0012.

Mueller, P. A., & Oppenheimer, D. M. (2014). The pen is mightier than the keyboard: Advantages of longhand over laptop note taking. *Psychological Science, 25*(6), 1159–1168.

Newen, A., De Bruin, L., & Gallagher, S. (2018). *The Oxford handbook of 4E cognition.* Oxford: Oxford Library of Psychology.

Norman, D. (1993). *Things that make us smart: Defending human attributes in the age of the machine.* New York: Basic Books.

Nyholm, S. (2023). *Artificial intelligence and human enhancement: Can AI technologies make us more (artificially) intelligent? Cambridge Quarterly of Health Ethics,* 1–13. https://doi.org/10.1017/S0963180123000464.

Peters, U. (2022). Reclaiming control: Extended mindreading and the tracking of digital footprints. *Social Epistemology, 36*(3), 267–282. https://doi.org/10.1080/02691728.2021.2020366.

Piredda, G., & Di Francesco, M. (2020). Overcoming the past-endorsement criterion: Toward a transparency-based mark of the mental. *Frontiers of Psychology, 11,* 1278. https://doi.org/10.3389/fpsyg.2020.01278.

Preston, B. (2013). *A philosophy of material culture: Action, function, and mind.* New York: Routledge.

Putnam, H. (1975). *Mind language and reality.* Cambridge: Cambridge University Press.

Rowlands, M. (2009). Extended cognition and the mark of the cognitive. *Philosophical Psychology, 22,* 1–19.

Rowlands, M. (2010). *New science of the mind: From extended mind to embodied phenomenology.* Cambridge, MA: The MIT Press.

Rupert, R. (2004). Challenges to the hypothesis of extended cognition. *Journal of Philosophy, 101,* 389–428.

Rupert, R. (2009). *Cognitive systems and the extended mind.* Oxford: Oxford University Press.

Sachs, B. (2011). The status of moral status. *Pacifc Philosophical Quarterly, 92,* 87–104.

Schettler, A., Raja, V., & Anderson, M. L. (2019). The embodiment of objects: Review analysis, and future directions. *Frontiers in Neuroscience, 13,* 1332.

Shapiro, L. (2014). *The Routledge handbook of embodied cognition*. Routledge, ⟨https://doi.org/10.4324/9781315775845⟩.

Smart, P. R., Andrada, G., & Clowes, R. W. (2022). Phenomenal transparency and the extended mind. *Synthese, 200*, 335. https://doi.org/10.1007/s11229-022-03824-6.

Søraker, J. H. (2007). The moral status of information and information technology: A relational theory of moral status. In S. Hongladarom, & C. Ess (Eds.). *Information technology ethics: Cultural perspectives* (pp. 1–19). Hershey: Idea Group Publishing.

Sterelny, K. (2004). Externalism, epistemic artefacts and the extended mind. In Richard Schantz, (Ed.). *The Externalist Challenge* (pp. 239–255). Berlin: Walter de Gruyter.

Sterelny, K. (2010). Minds: Extended or scaffolded? *Phenomenology and the Cognitive Sciences, 9*(4), 465–481.

Sutton, J. (2006). Distributed cognition: Domains and dimensions. *Pragmatics and Cognition, 14*(2), 235–247.

Sutton, J. (2010). Exograms and interdisciplinarity: History, the extended mind and the civilizing process. In R. Menary (Ed.). *The extended mind* (pp. 189–225). Cambridge: MIT Press.

Thomasson, A. (2007). Artifacts and human concepts. In E. Margolis, & S. Laurence (Eds.). *Creations of the mind: Theories of artifacts and their representation* (pp. 52–73). Oxford: Oxford University Press.

Valeriani, D., Santoro, F., & Ienca, M. (2022). The present and future of neural interfaces. *Frontiers in Neurorobotics, 16*, 953968. https://doi.org/10.3389/fnbot.2022.953968.

Varga, S. (2017). Demarcating the realm of cognition. *Journal for General Philosophy of Science, 49*, 435–450.

Vold, K. (2015). The parity argument for extended consciousness. *Journal of Consciousness Studies, 22*, 16–33.

Vold, K. (2018). Overcoming deadlock: Scientifc and ethical reasons to embrace the extended mind thesis. *Philosophy and Society, 29*(4), 471–646.

Vold, K., & Hernández-Orallo, J. (2021). AI extenders and the ethics of mental health. In F. Jotterand, & M. Ienca (Eds.). *Artificial intelligence in brain and mental health: Philosophical, ethical & policy issues. advances in neuroethics*. Cham: Springer, ⟨https://doi.org/10.1007/978-3-030-74188-4_12⟩.

Walter, S. (2010). Cognitive extension: The parity argument, functionalism, and the mark of the cognitive. *Synthese, 177*, 285–300.

Weiskopf, D. A. (2008). Patrolling the Mind's boundaries. *Erkenntnis, 68*, 265–276.

Wheeler, M. (2010). In defense of extended functionalism. In R. Menary (Ed.). *The extended mind* (pp. 245–270). Cambridge MA: MIT Press.

Wheeler, M. (2011). In search of clarity about parity: In Symposium on Andy Clark's supersizing the mind. *Philosophical Studies, 152*(3), 417–425.

Wheeler, M. (2016). A tale of two dilemmas: Cognitive kinds and the extended mind. In C. Kendig (Vol. Ed.), *Natural kinds and classification in scientific practice: 2015*, (pp. 210–221). London: Routledge.

Wheeler, M. (2019). Breaking the waves: Beyond parity and complementarity in the arguments for extended cognition. In M. Colombo, E. Irvine, & M. Stapleton (Eds.). *2019: Andy Clark and his critics*. Oxford: Oxford University Press.

Zilio, F. (2020). Extended mind and the brain-computer interface. A pluralist approach to human-computer integration. *Rivista Internazionale di Filosfoia e Psicologia, 11*(2), 169–189. https://doi.org/10.4453/rifp.2020.0011.

CHAPTER TEN

The ethical implications of indicators of consciousness in artificial systems

Michele Farisco[a,b,*]
[a]Centre for Research Ethics and Bioethics, Uppsala University, Sweden
[b]Biogem, Biology and Molecular Genetics Research Institute, Ariano Irpino, Italy
[*]Corresponding author. e-mail address: michele.farisco@crb.uu.se

Contents

1.	Introduction	185
2.	The prism of consciousness	187
3.	Indicators of consciousness and their application to AI	190
4.	Ethical implications of indicators of consciousness in AI	193
5.	Conclusion	196
	Acknowledgments	196
	References	196

Abstract

The prospect of artificial consciousness raises theoretical, technical and ethical challenges which converge on the core issue of how to eventually identify and characterize it. In order to provide an answer to this question, I propose to start from a theoretical reflection about the meaning and main characteristics of consciousness. On the basis of this conceptual clarification it is then possible to think about relevant empirical indicators (i.e. features that facilitate the attribution of consciousness to the system considered) and identify key ethical implications that arise. In this chapter, I further elaborate previous work on the topic, presenting a list of candidate indicators of consciousness in artificial systems and introducing an ethical reflection about their potential implications. Specifically, I focus on two main ethical issues: the conditions for considering an artificial system as a moral subject; and the need for a non-anthropocentric approach in reflecting about the science and the ethics of artificial consciousness.

1. Introduction

The prospect of artificial forms of consciousness is increasingly gaining traction as a concrete possibility both in the minds of lay people and

Developments in Neuroethics and Bioethics, Volume 7
ISSN 2589-2959, https://doi.org/10.1016/bs.dnb.2024.02.009
Copyright © 2024 Elsevier Inc. All rights are reserved, including those for text and data mining, AI training, and similar technologies.

of researchers in the field of neuroscience, robotics, AI, and their intersection (Butlin et al., 2023; LeDoux et al., 2023). While in the past the idea that an artificial system, like humanoid robots or non-anthropomorphic artefacts capable of emulating human cognitive and interactive features, can instantiate a form of consciousness was mainly promoted by sci-fi literature and movies, today it has much more scientific credit, notwithstanding significant criticisms still raised against it (Dietrich, Fields, Sullins, Van Heuveln, & Zebrowski, 2021). Even if eventually not accepted as a real possibility, today the idea of artificial consciousness is considered worthy of scientific discussion more than in the past. This even goes to the point that, for instance, the European Innovation Council has published a call dedicated to the topic of artificial awareness (EIC, 2021) selecting eight multidisciplinary research projects, including important contribution from neuroscience and AI, in order to make progress in theoretical, technological, and ethical research.

The reasons for this change in the perception of the scientific relevance of artificial consciousness are still to be clarified. My intuition is that three factors play an important role: the significant advancement in understanding the cerebral underpinnings of conscious perception, the impressive achievements of AI technologies, and the increasing interaction between neuroscience and AI. The combination of these three factors resulted, for instance, in brain-inspired AI, that is a form of AI informed by the knowledge of the brain (Farisco et al., 2023; Hassabis, Kumaran, Summerfield, & Botvinick, 2017; Summerfield, 2023).

Amid the ongoing debate about the conceptual plausibility and the technical feasibility of artificial forms of consciousness, there are some risks that demand careful consideration. At the theoretical level, there is first the need to clarify what we mean by consciousness. The definition of consciousness is the object of a hot debate and consensus seems very hard to achieve, notwithstanding some recent relevant attempts (Michel et al., 2019; Northoff & Lamme, 2020; Seth & Bayne, 2022; Wiese, 2020; Yaron, Melloni, Pitts, & Mudrik, 2022). Indeed, the lack of a shared definition of consciousness impacts the discussion about its artificial implementation, emulation, or replication, but I argue against overemphasizing the importance of a shared definition of consciousness in order to avoid the risk of *impasse*. Since it is likely impossible to get a general consensus on its definition, the debate should start from a technical and stipulative definition, and then strive for identifying some features or dimensions that everybody or the majority of people tend to attribute to consciousness. In a next step, one should try to identify some indicators, that are features or abilities that assist us in attributing consciousness to natural or artificial systems (Pennartz, Farisco, & Evers, 2019). In this way the big challenge of getting a shared definition of consciousness is replaced by a more pragmatic attempt to identify some relevant features that likely indicate the capacity of the system in question to implement conscious behavior. This approach, that resembles what in consciousness studies has recently been qualified as the real problem of consciousness (Seth, 2021) as opposed to the hard problem of consciousness (Chalmers, 1995), may appear too modest at a first glance, but I think that it is the only reasonable strategy to make concrete progress in the discussion about artificial consciousness, particularly to avoid the traps of conceptual *impasse* and linguistic incommunicability.

In what follows, I start with a reflection on the different dimensions of consciousness. Then, I describe the indicators that I have recently contributed to introducing, including a reference to their potential application to AI. Finally, I introduce an analysis of the potential ethical implications of the indicators of consciousness applied to AI.

2. The prism of consciousness

As already mentioned, reaching a consensus on how to define consciousness is a most challenging task. In fact, there is no agreement on what constitutes the specific characteristics of conscious experience, in other words, what distinguishes it from non-conscious cognitive activity.

Consciousness may assume different specific meanings in contemporary scientific and philosophical research. A distinction that has important clinical valence is that between the level or state of consciousness, that is wakefulness or sleep, and its content, for instance images or words (Laureys, 2005), even if levels and contents are arguably two distinct but not completely dissociable dimensions (Bachmann & Hudetz, 2014; Mashour, Roelfsema, Changeux, & Dehaene, 2020). Block introduced another popular distinction, even if it is not unanimously accepted (Naccache, 2018), differentiating between phenomenal consciousness (i.e. subjective experience or "what it is like to be") from access consciousness (i.e. availability of information for use in reasoning and rationally guiding speech and action) (Block, 1995).

These two conceptual distinctions suffice to outline the need for linguistic clarification when referring to consciousness. Specifically, the term consciousness should not be used without specifying it, for instance whether it refers to conscious processing of specific contents, to conscious access or to the state of consciousness (Farisco & Changeux, 2023). In the case of attempts to replicate consciousness artificially, it would be wise to specify which form of consciousness is the specific target. It may be that an artificial system is more likely able to emulate access consciousness rather than phenomenal consciousness (provided that this distinction is valid in the first place). Therefore, the conceptual distinction between the two forms of consciousness does not only have a theoretical valence, but it can also delimit the space of possibility of artificial consciousness.

In addition to the different forms of consciousness, its various dimensions need to be specified as well. The notion of a consciousness' dimension is particularly useful for a meaningful and balanced discussion about its artificial replication. Taking consciousness as a general reference is too vague for advancing in the clarification of the theoretical plausibility and the technical feasibility of contemporary attempts to replicate it, while focusing on some specific dimensions may allow us to get a more realistic view of what is logically conceivable and technically possible. The notion of consciousness' dimensions has been widely analyzed in a 2016 paper by Bayne, Hohwy, and Owen (Bayne, Hohwy, & Owen, 2016). Focusing on global states of consciousness (i.e. states of consciousness characterizing the overall conscious condition of a subject), they argue that consciousness is manifested in multiple ways in different global states, and that the notion of levels should be replaced by that of dimensions of consciousness to properly describe its multifaceted nature. The central thesis is that global states of consciousness are not gradable along one dimension, but rather distinguished along different dimensions. The authors introduce two main families of consciousness dimensions: content-related and functional. The first family includes, for instance, gating of conscious content, distinguishing low-level features from high-level features of an object. The second family includes, for instance, cognitive and behavioral control, that is the availability of conscious contents for control of thought and action. More specifically, Walter has recently proposed three content-related dimensions, namely sensory richness, high-order object representation, and semantic comprehension, and seven functional dimensions, comprising executive functioning, memory consolidation, intentional agency, reasoning, attention control, vigilance, and meta-awareness (Walter, 2021).

Also Chalmers has recently referred to different consciousness dimensions in a reflection about the relationship between Large Language Models

(LLM) and consciousness (Chalmers, 2023). Referring to conscious experience, he identifies five dimensions:

Further relevant reflections about the dimensions of consciousness come from Birch et al., who with reference to animal consciousness introduce five dimensions (J. Birch, Schnell, & Clayton, 2020), namely:

- *Perceptual-Richness* acknowledges that any measure is specific to a sense modality, so there is no overall level of perceptual richness. Additionally, within a particular sense modality, perceptual richness can be dissected into different components.
- *Evaluative-Richness* pertains to positive or negative affective valence, providing a foundation for decision-making. Similar to perceptual richness, also evaluative richness can be broken down into different components.
- *Unity* describes integration of experience at a time. Conscious experience is usually highly unified.
- *Temporality* describes the integration of experience across time. Conscious experience commonly takes the form of a continuous stream.
- *Self-consciousness* or *Selfhood* refers to awareness of oneself as distinct from the world outside.
- *Sensory experience:* e.g. seeing red;
- *Affective experience:* e.g. feeling pain;
- *Cognitive experience:* e.g. thinking hard;
- *Agentive experience:* e.g. *deciding to act;*
- *Self-consciousness:* i.e. awareness of oneself.

Therefore, while the detailed identification of further dimensions within the abovementioned two families, as well as the identification of other possible families of consciousness' dimensions remain an open issue (Veit, 2022; Walter, 2021), the concept of consciousness' profiles emerges as spaces of experience delimited by distinct dimensions. Accordingly, we can distinguish between consciousness' profiles not in terms of their overall levels along a single dimension, but rather with reference to the combination of different dimensions characterizing them. For instance, it may well be the case that the consciousness' profile of a human subject exhibits more advanced content-related and functional dimensions (e.g. semantic comprehension and meta-awareness, respectively) compared to a non-human entity. In contrast, other content-related and functional dimensions (e.g. sensory richness and vigilance, respectively) may be less advanced in some non-human beings. This does not imply that one overall conscious state is higher or lower than the other, but rather that it is shaped differently. Therefore, the comparison between human, other animals, and potential artificial consciousnesses should be framed in terms of resemblances and differences along specific dimensions rather than in terms of higher or lower levels along only one dimension. The key point is that consciousness is a multifaceted reality, akin to a prism, and it is irreducible to a single level of description.

This multidimensional view of consciousness is particularly beneficial for a balanced analysis of hypothetical artificial consciousness. It may be the case that some content-related and/or functional (or even other families of) consciousness' dimensions can be eventually emulated by artificial systems, defining a particular consciousness' profile. This profile may be less advanced than the human with reference to some dimensions (e.g. semantic comprehension and meta-awareness), while more advanced on other dimensions (e.g. sensory richness and memory consolidation). Based on a multidimensional perspective of consciousness, the prospect of artificial consciousness appears less implausible, at least in principle. The question then shifts to how different artificial consciousness' profiles may be from human consciousness. For instance, artificial consciousness' profiles may lack a sufficient level of certain dimensions considered crucial to be regarded as moral patients or subjects or both.

3. Indicators of consciousness and their application to AI

As mentioned above, advancing the discussion about artificial consciousness would greatly benefit from the introduction of a list of indicators that aid in identifying consciousness in both biological and non-biological entities. Methodologically, there are two viable options for contemplating indicators of consciousness: one can either rely on selected scientific theories of consciousness and infer the indicators from them, endorsing a theory-heavy approach (Butlin et al., 2023), or strive to remain neutral about any specific theory of consciousness and instead focus on features generally considered characteristic of it, adopting a theory-light approach (Birch, 2022).

Together with a neuroscientist and a philosopher, I have contributed to the formulation of six indicators of consciousness aimed at identifying it in other humans, in non-human animals, and potentially in artificial systems (Farisco, Pennartz, Annen, Cecconi, & Evers, 2022; Pennartz et al., 2019).

As highlighted in the introduction, the approach we decided to use aligns rather closely with the theory-light approach. We started from a very general, operational definition of consciousness as a multimodal situational survey, that is the construction and intentional use of world models, to subsequently identify its characteristics and infer relevant indicators.

More specifically, we assumed that consciousness takes the form of an inferential dynamic model or representation of the world, enabling complex decision-making and behavior directed towards relevant goals (Pennartz, 2015). Complex decision-making is a kind of process which requires various high-level intentional actions, in contrast to reflexes and habits, which can be largely non-conscious.

Based on this premise, we identified the following main characteristics of consciousness. While we think that these characteristics sufficiently qualify consciousness, we do not exclude the possibility that consciousness has also other features:

1. *Qualitative richness*: Conscious experience is qualified by distinct sensory modalities and submodalities. For example, in vision, submodalities include texture, motion, color, size, shape, and depth.
2. *Situatedness*: Consciousness is specified by the particular spatiotemporal condition of the subject, whose body occupies a particular place in space and time. Importantly, this concept includes objects with specific spatiotemporal relationships to each other (rather than departing from abstract space or time itself), as well as the subject's body (as one object).
3. *Intentionality*: Consciousness is about something other than its neuronal underpinnings.
4. *Integration*: The components of the conscious experience are perceived as a unified whole.
5. *Dynamics and stability*: Conscious experiences include both dynamic changes and short-term stabilization.

Building upon the above definition and characteristics of consciousness, we introduced the following indicators, encompassing different types and including behavioral, neurophysiological, and cognitive dimensions:

1. *Goal directed behavior (GDB) and model-based learning*: GDB is technically defined as behavior driven by the expected consequences of an action, and where the agent knows that their action is causally linked to obtaining a desirable outcome. Model-based learning depends on the agent's ability to have an explicit model of themselves and their surrounding world.

2. *Brain anatomy and physiology*: Since the consciousness of mammals depends on the integrity of particular cerebral systems (i.e. thalamo-cortical systems), it is reasonable to think that similar structures or homologous functions indicate the presence of consciousness.
3. *Psychometrics and meta-cognitive judgment*: If the agent can detect and discriminate stimuli and make some meta-cognitive judgments about perceived stimuli, they are probably conscious.
4. *Episodic memory*: If the agent can remember events (what) experienced at a particular place (where) and time (when), they are probably conscious.
5. *Acting out one's subjective, situational survey*: If an agent is susceptible to illusions and perceptual ambiguity, they are probably conscious.
6. *Acting out one's subjective, situational survey: visuospatial behavior.* Our last proposed indicator of consciousness is the ability to perceive objects as stably positioned, even when the agent moves in their environment and scans it with their eyes.

This list is inspired by a pragmatic *epistemic humility*. It is conceived to be provisional and heuristic yet at the same time operational. Importantly, the indicators serve as *positive* markers of consciousness (Ginsburg & Jablonka, 2019): While their presence suggests the presence of consciousness, their absence does not rule out the possibility of undetected consciousness. In the case of AI systems, potential additional and more specific indicators may take the form of ecological and social or cooperative indicators. More specifically, if an AI system displays the ability of coordinating functions with other systems to maximize the chances of goal achievement, then it may be conscious. The likelihood that an AI system capable of this kind of behavior is conscious also increases if it can postpone satisfying its needs for a delayed, more rewarding outcome, or if it can anticipate the emergence of future needs.

While the list of indictors reported above is sufficiently concrete to help identifying consciousness in both biological organisms and artificial systems, it is no definitive answer to the challenge of detecting and characterizing consciousness in other entities. In fact, the indicators are clearly affected by a number of limitations, including a limited focus on possible basic, non-cognitive forms of consciousness, such as affective-based conscious experiences (Solms, 2021; Vandekerckhove, Bulnes, & Panksepp, 2014). The taxonomy introduced by Tulving is particularly useful in order to discriminate between more basic and more sophisticated conscious abilities (Tulving, 2005). According to him, there are three kinds of consciousness: autonoetic, which is relative to the knowledge of the self; noetic, which is relative to the knowledge of the outside world; and anoetic, which is relative to the absence of explicit actual knowledge (LeDoux & Lau, 2020). Anoetic consciousness is conceived as the condition to be alive and responsive to stimuli as opposed to having explicit conscious contents (LeDoux, 2021), or as "a stream of pre-reflective affective and sensorial perceptual consciousness essential for the waking state of the organism in the absence of an explicit self-referential awareness of associated cognitive contents". Therefore, the anoetic form of consciousness is conceived as the most basic, serving as an essential condition for the emergence of the other two: while it is theoretically possible to have anoetic consciousness without noetic or autonoetic consciousness, the opposite is not possible. This point is intuitively true for biological organisms, but it raises a peculiar challenge for artificial systems like LLM, which are instructed through high level commands while apparently lacking the more basic anoetic form. Provided that AI can actually be conscious, can it have a more sophisticated form of consciousness (i.e. noetic or autonoetic) in the absence of the fundamental form of anoetic consciousness? And, if so, how can this be understood and explained (LeDoux et al., 2023)? The challenge here is both theoretical and practical. At the theoretical level, one should avoid the risk of reasoning anthropocentrically, eventually excluding *a-priori* two possibilities based on anthropocentric and anthropomorphic biases: One cannot off-hand dismiss the potential of AI to develop either more sophisticated forms of consciousness even if lacking anoetic consciousness or some forms of anoetic consciousness alternative to those experienced by biological organisms, for instance if connected to robotic systems.

From these considerations it follows that, at the practical level, we should strive to identify indicators of consciousness that are not limited to noetic and autonoetic consciousness, but refer also to anoetic consciousness, without considering the last an entailment of the former. This adds a further element of complexity to the already demanding challenge of identifying indicators of consciousness in artificial systems.

4. Ethical implications of indicators of consciousness in AI

In addition to theoretical issues of conceivability and plausibility, and to technical issues of feasibility, the prospect of artificial consciousness also raises a number of ethical issues (Hildt, 2023). The fundamental ethical question is whether we should strive to build an artificial form of consciousness, and if so, why, for what reason and for what goal. Opinions diverge on this point. Notably, Metzinger calls for an international moratorium against any attempt to build artificial consciousness because this kind of research may end up in inadvertently giving rise to synthetic phenomenology and possibly to artificial suffering (Metzinger, 2021). The position argued for by Metzinger is quite strong, and it is inspired by the moral imperative to minimize any form of suffering, including both biological and non-biological suffering: The risk to cause the latter is sufficient to stop research towards artificial consciousness. A somehow complementary position is stopping research on AI consciousness because of the risk of losing control over AI by humans, as recently stressed by Marcus, among others (Marcus, 2023). The argument questions why we should open another, likely riskier box, when society is not able to really manage the impact of LLM on our society. Precaution would therefore seem ethically reasonable. Yet calling for pausing or even stopping research on AI is questionable, as recent debate on LLM showed (FLI, 2023; Ienca, 2023). More pragmatically, I think that a more useful ethical approach to the issue of conscious AI would be to start from indicators of consciousness like those listed above and reflect on their potential ethical consequences. For instance, GDB and model-based learning are indicative of the ability to have conscious interests, to recognize the relevance of external inputs for fulfilling those interests, and to act on those inputs accordingly. If present in AI systems, these indicators are evidence of ethically salient features, arguably making the system an ethical subject as well as an ethical patient.

Concerning brain anatomy and physiology, despite the absence of analogous or homologous anatomical and physiological structures in artificial systems, it is theoretically possible to artificially implement equivalent functionalities. While these are necessary for consciousness, their sufficiency for it is questionable, and depends on the background theoretical framework. For instance, some researchers recently reflected on the technical feasibility of artificial consciousness assuming as starting point the validity of computational functionalism, that is that consciousness does not depend on the underlying structure (Butlin et al., 2023). This revival of an old theory that has long been considered conclusively rejected by prominent neuroscientists (Changeux, 1986; Edelman, 1992; Pennartz, Forthcoming) may seem a dead end, but its attraction has never quite vanished in some circles. For those who take it seriously, it is both theoretically and ethically relevant. For instance, it would suggest a form of ontological dualism, which inspire anthropological models that do not recognize the relevance of bodily dimensions for human identity and action. Ethical considerations therefore also suggest caution on this point.

In general, for all the indicators above, a fundamental ethical question is how to test them in AI systems. In fact, these indicators have been introduced with reference to biological organisms, primarily to human subjects, and their applicability to AI may be limited both because they may not be relevant to AI, and because AI may show some forms or levels of consciousness too different from those resulting from biological evolution. Consequently, one should avoid falling into the traps of anthropomorphic or anthropocentric interpretations (Salles, Evers, & Farisco, 2020).

A final point worth considering is the possible relevance of the above indicators of consciousness to the debate about the possibility of AI being a moral subject. According to some authors, phenomenal consciousness is considered both necessary and sufficient for attributing moral status (Levy, 2014), that is, for considering an entity as deserving moral consideration, even though this position is object of debate (Farisco & Evers, 2017; Shepherd, 2023). Technically speaking, an entity that has phenomenal experience, positive or negative, is defined in moral philosophy as a moral patient, that is as the object of moral consideration and respect. A related important concept in moral philosophy is moral subjectivity: an entity is a moral subject if it has the ability to make moral decisions. If so, it can be considered as accountable and responsible of what it decides to do and/or not to do.

Can an AI system be considered as a moral patient, or subject, or both? How can the indicators of consciousness introduced above help in clarifying these issues? Generally speaking, I think that for qualifying an entity as a moral patient, phenomenal consciousness is not necessary, even if sufficient. For instance, we may consider AI as deserving moral respect not for intrinsic reasons such as phenomenal consciousness but for extrinsic reasons, for instance because it considerably improves the well-being of people, and it would therefore be morally wrong to damage or destroy it. Even if consciousness is not generally necessary for being qualified as a moral patient, it may provide additional reasons and justification for it though, and can also be considered a sufficient condition for an entity to be a moral patient. In comparison, moral subjectivity appears more dependent on consciousness: capacity for consciousness, including phenomenal consciousness, is arguably a necessary condition for an entity to be considered as a moral subject. It may be debated whether consciousness is sufficient for moral subjectivity, for instance since additional cognitive and evaluative capacities may be necessary for moral accountability and responsibility. Yet, the fact that an entity should be conscious to be considered as a moral subject appears intuitively reasonable: While it remains an open question what kind of consciousness, for instance whether access consciousness or phenomenal consciousness, is necessary for moral subjectivity, a minimal capacity for subjective evaluation seems crucial.

As mentioned, the ethical debate about potential conscious AI is open and characterized by significant controversies (Hildt, 2023), which go beyond the limited space of this chapter. Do the abovementioned indicators of consciousness help clarifying whether an AI system can be considered a moral subject though? In principle, the answer seems yes: since they are conceived to assist in attributing consciousness, and since consciousness is arguably a necessary condition for moral subjectivity, the indicators have the potential to provide an important contribution to the debate about hypothetical artificial moral subjectivity. Yet there is an important limitation to highlight, which deserves more attention in the future: the indicators here described do not directly refer to phenomenal consciousness. Even if considered not strictly necessary for moral subjectivity, phenomenal consciousness arguably plays an important role for it. Therefore, it is necessary to further expand the reflection about the indicators of consciousness taking phenomenal experience as a specific target.

5. Conclusion

In order to advance in the ethical debate about artificial consciousness, I propose to start from some features of consciousness in order to identify relevant indicators. In this chapter I presented some indicators previously introduced with reference to human and non-human animals, stressing their potential relevance also to AI. Further reflection is needed to explore the connection of these and other indicators with phenomenal consciousness, particularly to advance in the debate about the possibility that AI may be considered as a moral subject.

Acknowledgments

The research is supported by funding from the European Union's Horizon 2020 Framework Programme for Research and Innovation under the Specific Grant Agreement No. 945539 (Human Brain Project SGA3) and by the European Innovation Council Action Acronym CAVAA, No. 101071178. I thank Kathinka Evers and Mehdi Khamassi for extremely inspiring discussions on the topic of this chapter.

References

Bachmann, T., & Hudetz, A. G. (2014). It is time to combine the two main traditions in the research on the neural correlates of consciousness: $C = L \times D$. *Frontiers in Psychology, 5*, 940. https://doi.org/10.3389/fpsyg.2014.00940.

Bayne, T., Hohwy, J., & Owen, A. M. (2016). Are there levels of consciousness? *Trends in Cognitive Sciences, 20*(6), 405–413. https://doi.org/10.1016/j.tics.2016.03.009.

Birch (2022). The search for invertebrate consciousness. *Noûs, 56*(1), 133–153. https://doi.org/10.1111/nous.12351.

Birch, J., Schnell, A. K., & Clayton, N. S. (2020). Dimensions of animal consciousness. *Trends in Cognitive Sciences, 24*(10), 789–801. https://doi.org/10.1016/j.tics.2020.07.007.

Block, N. (1995). On a confusion about a function of consciousness. *Behavioral and Brain Sciences, 18*(2), 227–287.

Butlin, P., Long, R., Elmoznino, E., Bengio, Y., Birch, J., Constant, A., & VanRullen, R. (2023). *Consciousness in artificial intelligence: Insights from the science of consciousness.* Retrieved from arXiv:2308.08708v3.

Chalmers, D. (1995). Facing up to the hard problem of consciousness. *Journal of Consciousness Studies, 2*, 200–219.

Chalmers, D. (2023). Could a large language model be conscious? Retrieved from arXiv:2303.07103 website.

Changeux, J. P. (1986). *Neuronal man: The biology of mind.* New York: Oxford University Press.

Dietrich, E., Fields, C., Sullins, J. P., Van Heuveln, B., & Zebrowski, R. (2021). *Great philosophical objections to artificial intelligence: The history and legacy of the AI wars,* 1.

Edelman, G. M. (1992). *Bright air, brilliant fire: On the matter of the mind.* New York: BasicBooks.

EIC. (2021). EIC Pathfinder challenge: Awareness inside. ⟨https://eic.ec.europa.eu/eic-funding-opportunities/calls-proposals/eic-pathfinder-challenge-awareness-inside_en⟩.

Farisco, M., Baldassarre, G., Cartoni, E., Leach, A., Petrovici, M. A., Rosemann, A., & Albada, S. J. (2023). A method for the ethical analysis of brain-inspired AI. Retrieved from arXiv:2305.10938v1 website.

Farisco, M., & Changeux, J. P. (2023). About the compatibility between the perturbational complexity index and the global neuronal workspace theory of consciousness. *Neuroscience of Consciousness, 2023*(1), niad016. https://doi.org/10.1093/nc/niad016.

Farisco, M., & Evers, K. (2017). The ethical relevance of the unconscious. *Philosophy, Ethics, and Humanities in Medicine: PEHM, 12*(1), 11. https://doi.org/10.1186/s13010-017-0053-9.

Farisco, M., Pennartz, C., Annen, J., Cecconi, B., & Evers, K. (2022). Indicators and criteria of consciousness: Ethical implications for the care of behaviourally unresponsive patients. *BMC Medical Ethics, 23*(1), https://doi.org/10.1186/s12910-022-00770-3.

FLI. (2023). Pause giant AI experiments: An open letter. Retrieved from ⟨https://futureoflife.org/open-letter/pause-giant-ai-experiments/⟩.

Ginsburg, S., & Jablonka, E. (2019). *The evolution of the sensitive soul: Learning and the origins of consciousness.* Cambridge, Massachusetts: The MIT Press.

Hassabis, D., Kumaran, D., Summerfield, C., & Botvinick, M. (2017). Neuroscience-inspired artificial intelligence. *Neuron, 95*(2), 245–258. https://doi.org/10.1016/j.neuron.2017.06.011.

Hildt, E. (2023). The prospects of artificial consciousness: Ethical dimensions and concerns. *AJOB Neuroscience, 14*(2), 58–71. https://doi.org/10.1080/21507740.2022.2148773.

Ienca, M. (2023). Don't pause giant AI for the wrong reasons. *Nature Machine Intelligence.* https://doi.org/10.1038/s42256-023-00649-x.

Laureys, S. (2005). The neural correlate of (un)awareness: Lessons from the vegetative state. *Trends in Cognitive Sciences, 9*(12), 556–559. https://doi.org/10.1016/j.tics.2005.10.010.

LeDoux, Birch, J., Andrews, K., Clayton, N. S., Daw, N. D., Frith, C., & Vandekerckhove, M. M. P. (2023). Consciousness beyond the human case. *Current Biology, 33*(16), R832–R840. https://doi.org/10.1016/j.cub.2023.06.067.

LeDoux, & Lau, H. (2020). Seeing consciousness through the lens of memory. *Current Biology: CB, 30*(18), R1018–R1022. https://doi.org/10.1016/j.cub.2020.08.008.

LeDoux, J. E. (2021). What emotions might be like in other animals. *Curr Biol, 31*(13), R824–R829.

Levy, N. (2014). The value of consciousness. *Journal of Consciousness Studies, 21*(1–2), 127–138.

Marcus, G. (2023). Sentient AI: For the love of Darwin, let's stop to think if we should. ⟨https://garymarcus.substack.com/p/sentient-ai-for-the-love-of-darwin⟩.

Mashour, G. A., Roelfsema, P., Changeux, J. P., & Dehaene, S. (2020). Conscious processing and the global neuronal workspace hypothesis. *Neuron, 105*(5), 776–798. https://doi.org/10.1016/j.neuron.2020.01.026.

Metzinger, T. (2021). An argument for a global moratorium on synthetic phenomenology. *Journal of Articial Intelligence and Consciousness, 8*(1), 1–24.

Michel, M., Beck, D., Block, N., Blumenfeld, H., Brown, R., Carmel, D., & Yoshida, M. (2019). Opportunities and challenges for a maturing science of consciousness. *Nature Human Behaviour, 3*(2), 104–107. https://doi.org/10.1038/s41562-019-0531-8.

Naccache, L. (2018). Why and how access consciousness can account for phenomenal consciousness. *Philosophical Transactions of the Royal Society of London. Series B, Biological Sciences, 373*(1755), https://doi.org/10.1098/rstb.2017.0357.

Northoff, G., & Lamme, V. (2020). Neural signs and mechanisms of consciousness: Is there a potential convergence of theories of consciousness in sight? *Neuroscience and Biobehavioral Reviews, 118*, 568–587. https://doi.org/10.1016/j.neubiorev.2020.07.019.

Pennartz (2015). *The brain's representational power. On consciousness and the integration of modalities.* Cambridge, MA: MIT Press.

Pennartz. (Forthcoming). *The riches of consciousness: How the brain creates our reality.* London: Taylor and Francis.

Pennartz, Farisco, M., & Evers, K. (2019). Indicators and criteria of consciousness in animals and intelligent machines: An inside-out approach. *Frontiers in Systems Neuroscience, 13*, 25. https://doi.org/10.3389/fnsys.2019.00025.

Salles, A., Evers, K., & Farisco, M. (2020). Anthropomorphism in AI. *AJOB Neuroscience, 11*(2), 88–95. https://doi.org/10.1080/21507740.2020.1740350.

Seth. (2021). *Being you the inside story of your inner universe (pp. 1 online resource).* ⟨http://link.overdrive.com/?websiteID=110056&titleID=5068666⟩.

Seth, A. K., & Bayne, T. (2022). Theories of consciousness. *Nature Reviews. Neuroscience, 23*(7), 439–452. https://doi.org/10.1038/s41583-022-00587-4.

Shepherd, J. (2023). Non-human moral status: Problems with phenomenal consciousness. *AJOB Neuroscience, 14*(2), 148–157. https://doi.org/10.1080/21507740.2022.2148770.

Solms, M. (2021). *The hidden spring: A journey to the source of consciousness* (First edition.,). New York, NY: W.W. Norton & Company.

Summerfield, C. (2023). *Natural general intelligence: How understanding the brain can help us build AI.* New York: Oxford University Press.

Tulving, E. (2005). Episodic memory and autonoesis: Uniquely human? In H. S. Terrace, & J. Metcalfe (Eds.). *The missing link in cognition* (pp. 4–56). New York: Oxford University Press.

Vandekerckhove, M., Bulnes, L. C., & Panksepp, J. (2014). The emergence of primary anoetic consciousness in episodic memory. *Frontiers in Behavioral Neuroscience, 7*, 210. https://doi.org/10.3389/fnbeh.2013.00210.

Veit, W. (2022). The origins of consciousness or the war of the five dimensions. *Biological Theory, 17*(4), 276–291. https://doi.org/10.1007/s13752-022-00408-y.

Walter, J. (2021). Consciousness as a multidimensional phenomenon: Implications for the assessment of disorders of consciousness. *Neuroscience of Consciousness, 2021*(2), niab047. https://doi.org/10.1093/nc/niab047.

Wiese, W. (2020). The science of consciousness does not need another theory, it needs a minimal unifying model. *Neuroscience of Consciousness, 2020*(1), niaa013. https://doi.org/10.1093/nc/niaa013.

Yaron, I., Melloni, L., Pitts, M., & Mudrik, L. (2022). The ConTraSt database for analysing and comparing empirical studies of consciousness theories. *Nature Human Behaviour, 6*(4), 593–604. https://doi.org/10.1038/s41562-021-01284-5.

CHAPTER ELEVEN

Moral dimensions of synthetic biological intelligence: Unravelling the ethics of neural integration

Masanori Kataoka[a], Christopher Gyngell[b,c], Julian Savulescu[b,d,e], and Tsutomu Sawai[a,e,f,*]

[a]Graduate School of Humanities and Social Sciences, Hiroshima University, Higashi–Hiroshima, Japan
[b]Murdoch Children's Research Institute, Melbourne, Australia
[c]Department of Paediatrics, University of Melbourne, Melbourne, Australia
[d]Faculty of Philosophy, University of Oxford, Oxford, Uinted Kingdom
[e]Centre for Biomedical Ethics, Yong Loo Lin School of Medicine, National University of Singapore, Singapore, Singapore
[f]Institute for the Advanced Study of Human Biology (ASHBi), Kyoto University, Kyoto, Japan
*Corresponding author. e-mail address: tstmsw@hiroshima-u.ac.jp

Contents

1. Introduction		200
2. Moral standing-related ethical dimensions of SBI systems		202
	2.1 Moral standing and human provenance	202
	2.2 Consciousness: A cornerstone of moral implications	203
	2.3 Agency and intelligence: Implications and ascription	204
	2.4 Utilisation and training: Ethical framework	206
3. Non moral standing-related ethical dimensions of SBI systems		207
	3.1 Consent	207
	3.2 Donor responsibility for the behaviour of SBI systems	207
	3.3 Commercialization and benefit sharing	209
4. Conclusion		210
Acknowledgements		211
Competing interests		211
References		211

Abstract

Recent research emphasizes the transformative potential of cultured human brain tissues in biocomputing applications. Employing neural cells derived from human induced pluripotent stem cells, a recent study interfaced these cells with multi-electrode arrays to craft a sophisticated real-time closed-loop feedback system the authors called 'DishBrain'. This innovation marks a new paradigm in synthetic biolo-gical intelligence (SBI) systems, integrating biological and silicon components in vitro

---

Developments in Neuroethics and Bioethics, Volume 7
ISSN 2589-2959, https://doi.org/10.1016/bs.dnb.2024.02.012
Copyright © 2024 Elsevier Inc. All rights reserved, including those for text and data mining, AI training, and similar technologies.

to exhibit goal-directed and intelligent behaviours. While SBI holds considerable promise, it also introduces unique ethical concerns, especially given its reliance on human biological material. We categorize these concerns into two groups: those related to the moral standing of SBI systems and those that are not. In the former category, SBI systems, due to their human neural components, may more readily exhibit properties indicative of moral standing compared to existing artificial intelligence systems. In the latter category, SBI technologies raise ethical dilemmas related to cell donors, encompassing issues of consent, donors' responsibility for behaviours of the SBI systems, and benefit sharing. Comprehensive ethical deliberation and empirical studies on these matters are imperative to ensure the responsible and socially acceptable evolution of SBI technologies.

1. Introduction

Recent research has highlighted the potential of cultured human brain tissues in biocomputing (Kagan et al., 2022). The study utilised neural cells derived from human induced pluripotent stem cells, integrating them with multi-electrode arrays to establish an intricate real-time closed-loop feedback system, which the authors called DishBrain. DishBrain demonstrated the capacity to 'learn' to play a simulated variant of the classic arcade table tennis game 'Pong', trained by either predictable or unpredictable stimuli depending on whether the stimulated ball was hit back or not. A key feature of DishBrain lies in its ability not only to passively adapt to the environment, but also to actively modify the environment by manipulating the simulated paddle within 'Pong'. In this sense, DishBrain manifests goal-directed behaviour.

The concept of biocomputing, involving the use of biological components for computational pursuits, is not new, and researchers have used both human and animal neural cells or tissues for such purposes. For instance, systems of cultured rat neurons interfaced with microelectrodes have been shown to selectively filter signals from noise (Isomura, Kotani, & Jimbo, 2015; for an overview, Kagan et al., 2023). What sets DishBrain apart is its capacity to adapt to environmental changes. Moreover, it integrates neural cells derived from human pluripotent stem cells, bridging the realms of stem cell research and artificial intelligence (AI).

DishBrain marks a new approach to developing synthetic biological intelligence (SBI) systems, amalgamating biological and silicon substrates in vitro to achieve goal-oriented or other intelligent behaviours (Kagan et al., 2023). Smirnova et al. (2023) proposed a similar but narrower conceptual framework called organoid intelligence (OI). Their framework focuses on intelligence facets such as learning, memory, and cognition and probes them in human brain organoids, understood as self-organized, three-dimensional brain tissue constructs derived from pluripotent stem cells. Despite shared theoretical underpinnings, DishBrain differs from OI due to its two-dimensional nature of the incorporated neural cells. In the subsequent ethical discussion, we focus on SBI utilizing human pluripotent stem cells, hereafter referred to simply as 'SBI'.

SBI technologies may offer advantages over conventional AI systems in terms of both energy consumption and learning efficiency (Smirnova et al., 2023). Human brains execute intricate calculations with less energy than AI systems do, signifying that integrating human neurons into SBI systems may markedly curtail computational energy requirements. Contemporary AI technologies, notably deep learning, are fraught with notoriously energy-intensive computational operations raising economic, environmental, and equity-related dilemmas (Strubell, Ganesh, & McCallum, 2019). SBI technologies may offer the potential to alleviate these concerns (Kagan et al., 2023). Moreover, current machine learning demands extensive data and time, whereas SBI can potentially leverage human-like computational abilities to expedite processes like pattern recognition, contributing to learning efficiency. DishBrain's ability to acquire game proficiency swiftly relative to that of AI systems—albeit without playing it better—exemplifies this efficiency (Le Page, 2021). Such learning efficiency extends beyond just saving time and enhancing computational processes. For example, it could make it easier to circumvent the persistent challenges of navigating intellectual property rights when collecting training datasets for machine learning. In particular, the reduced data requirements for an SBI system could lead to simpler and less contentious data collection processes.

The maturation of SBI systems holds promise for research on the human brain. Such systems may provide a better understanding of the computational dynamics inherent in human neural tissues, particularly their interplay with external milieu (Kagan et al., 2022). Moreover, these systems could provide models for neurodegenerative, neurodevelopmental, and mental disorders (Smirnova et al., 2023).

Despite its immense potential, SBI also introduces new ethical quandaries that warrant proactive scrutiny. As previously noted, SBI systems have potential advantages over AI systems across specific performance benchmarks. Consequently, the concerns currently surrounding AI technologies might become more relevant for SBI. However, we will not discuss ethical issues shared by SBI and AI, such as concerns related to AI- or SBI-induced unemployment or, more speculatively, SBI systems possessing intelligence surpassing that of humans. Our focus is on identifying and scrutinizing the ethical predicaments unique to SBI, arising from its utilisation of human biological materials, which distinctly demarcates it from existing AI technologies. To do so, we first explore issues centring on the moral standing of SBI systems, and then proceed to look at other ethical concerns including consent, donor responsibility and commercialization.

2. Moral standing-related ethical dimensions of SBI systems

2.1 Moral standing and human provenance

The attribution of moral standing to SBI systems hinges on fundamental questions regarding the nature of moral standing. Moral standing is intricately linked to many defining attributes, some of which may potentially be achievable by SBI systems in the future. These attributes encompass phenomena such as phenomenal consciousness, sentience (including the capacity to experience pleasure and pain), intelligence (expressed through goal-oriented behaviours), free will (embodied in reason-responsiveness or higher-order motivational dynamics), and moral understanding. This list of morally relevant capacities is adapted from the discussion of the moral standing of AI systems (Sinnott-Armstrong & Conitzer, 2021). In this regard, there is no difference between AI systems, SBI systems or any other entity: if they possess these capacities, they have moral standing.

However, SBI systems may have several features that could make them more likely to acquire moral standing than functionally equivalent AI systems. One candidate is their human cellular constituents. Some theories posit that there is something inherently special about human life, conferring humans with a high moral standing independent of their capacities. Following these accounts, humans always enjoy higher moral standing than animals, even if some humans have less capacities. Similarly, one might argue that the inclusion of human cells in SBI systems imparts them with a higher moral standing than less competent silicon-based systems. However, the claim that humans inherently have a high moral standing regardless of their capacities has faced rigorous criticism as epitomizing speciesism, that is, the view that we should treat members of our species better simply because of our biological connection with them. Nevertheless, a more nuanced perspective may be possible. In relation to human brain organoids, some argue that their human origin may be merely one contributing facet to determining their moral standing, along with other morally relevant features (Zilio & Lavazza, 2023; Zilio, 2023). If this approach were extrapolated to an SBI system, it would potentially be ascribed a higher moral standing than its AI counterpart.

2.2 Consciousness: A cornerstone of moral implications

One attribute commonly thought to underpin moral standing is 'consciousness'. The prospect of SBI systems acquiring consciousness in the future raises deep and complex ethical concerns. The implications of cultured human brain tissues acquiring consciousness have been the subject of much recent scholarly discourse in the context of brain organoid research (Koplin & Savulescu, 2019; Niikawa, et al., 2022). This concern currently rests largely within the realm of speculation, primarily due to the constraints present in in vitro brain organoids (Hyun et al., 2022). Critical dissimilarities in size, maturity, structural complexity, and the absence of functional interactions with the environment distinguish in vitro human brain organoids from human brains.

SBI technologies possess a greater potential to acquire consciousness than current brain organoid models and inorganic computers, owing to the interaction of biological and silicon compounds. Nerve cells are the only substrate known to support consciousness processes, and thus, inorganic computers may lack the inherent capacity to support phenomenal experience. While brain organoids are composed of nerve cells, current models lack systems for sensory input and feedback. In animals, most conscious experiences are linked to input from various sensory systems. In SBI, the silicon component can act like the sensory systems for the neural cells, providing a mechanism for the feedback of external stimuli to the internal state of the system. In DishBrain, the system is exposed to unpredictable stimuli when it gets a response wrong (the paddle misses the ball), and predictable stimuli when it gets it right. The experience of pleasure and pain are motivators of animal behaviour. If it was possible for an SBI system to experience wrong responses as pain, and right responses as pleasure, this may drive their learning process when being trained to learn something novel such as playing pong. Since an enhanced learning process constitutes an attractive design feature, developers may imbue SBI with the necessary prerequisites for perceiving pleasure and pain, and the environmental architecture to evolve this capacity—even if unbeknown to themselves. In fact, the capacity to incorporate sensory feedback is an advantage of SBI technologies, which could lead to a transformative shift in brain organoid research. To date, the exploration of potentially conscious brain organoids has largely been confined to in vitro constructs. However, it is imperative that future discussions incorporate the development of SBI technologies.

Nevertheless, it is crucial not to overestimate the environmental interactions of SBI systems. Neural structures known to support consciousness (in brains) are many times more complex and intricate than anything conceivably achievable with SBI systems in the short term. Confining cultured neural tissues within constrained spaces, as typified by the DishBrain configuration, may render attaining consciousness impossible owing to spatial constraints on the volume of neural tissues. While these limitations could potentially be circumvented through more mature and extensive neural tissues or parallelizing multiple subsystems within the SBI, expanding the system in this manner jeopardizes its functional utility as a computing entity. Therefore, the likelihood that SBI systems achieve consciousness in the short term is tenuous.

Overall, nuance and clarity when discussing capacities of SBI systems are crucial. Terminology related to human capacities in the context of SBI should not be assumed to have the same meaning across different contexts (Smirnova et al., 2023). A case in point is attributing 'sentience' to DishBrain (Kagan et al., 2022), where 'sentience' denotes responsiveness to sensory stimuli, without necessarily implying subjective experience (Friston, Wiese, & Hobson, 2020). This diverges from 'sentience' in bioethics, where it signifies the capacity for pleasure or pain, serving as a pivotal determinant of moral standing. Resembling the disarray resulting from conflating nociceptive responses and subjective pain experiences in the discourse surrounding the moral standing of the human foetus (Salomons & Iannetti, 2022), a non-nuanced grasp of the terminological landscape surrounding SBI can result in convoluted ethical assertions.

2.3 Agency and intelligence: Implications and ascription

Other approaches to moral standing stress the capacity for agency and intelligence rather than conscious experience. The term 'intelligent' aptly characterizes the behaviour of DishBrain, signifying its adeptness in goal-directed conduct, which can be seen as indicative of some form of intelligence. Consequently, an SBI system that adeptly demonstrates such goal-oriented behaviours can be attributed with a degree of agency. Indeed, DishBrain has sparked discussion surrounding the agency and subsequent moral standing of SBI systems. Shepherd (2023) posited that if an SBI system demonstrated activities contributing to its well-being—encompassing pursuits such as fulfilling desire, developing its inherent nature, cultivating relationships, achieving important goals, and acquiring knowledge—it could possess a modicum of moral standing, even if devoid of consciousness. This possibility has two implications. First, the agency demonstrated by an SBI system may not necessarily be equivalent to the agency of moral importance possessed by a human (Kataoka & Sawai, 2023); for instance, moral agency cannot be attributed to an arcade table implementing 'Pong'. In both AI and SBI domains, acquiring the necessary agency to warrant moral standing can be a protracted process; however, SBI technologies may potentially expedite this progression (Buchanan, 2018). Second, while future SBI systems may exhibit sufficiently intricate behaviours characteristic of moral agency, the same also applies to AI systems. Compared to the scepticism surrounding the consciousness of inorganic AI systems, scepticism surrounding their agency is minimal. Thus, issues around the moral agency of SBI systems can be considered as a natural extension of those of AI systems.

Ethical concerns that are more specific to SBI technologies, especially regarding agency, appear to be intertwined with the agency attributed by human actors. Psychological studies have affirmed that humans tend to ascribe agency to AI systems (Gray, Gray, & Wegner, 2007). The same tendency would likely be observed in relation to SBI systems, perhaps to an even greater degree, due to the incorporation of human biological constituents, although empirical investigations into how SBI systems are perceived by people has yet to begin.

Attributing agency to SBI systems that exceeds their actual capabilities can lead to a myriad of ethical and societal problems, including public resistance to their use if SBI technologies are believed to manipulate entities of moral significance (Smirnova et al., 2023). This could result in an overregulation of research on these systems and produce far-reaching repercussions for human well-being, especially considering the substantive potential of SBI technologies for medical applications. Hence, we underscore the necessity for future empirical studies on people's perceptions of SBI systems coupled with judicious science communication for the robust evolution of SBI technologies. A significant impediment to this pursuit will be mitigating the effects of sensationalism and the prevalence of hyperbole on SBI technologies. This concern has surfaced in the context of AI technologies and research on human brain organoids (Kataoka et al., 2023; Leufer, 2020); these mistakes should not be repeated in the adoption of SBI technologies. Although there is still a significant developmental path till SBI technologies attain moral standing, the issue of agency attribution is urgent and should be prioritized and addressed.

2.4 Utilisation and training: Ethical framework

Even if SBI systems eventually attain moral standing, it does not necessarily mean that their creation or research use should be prohibited. Instead, well-considered regulations are required to address this eventual reality. The ethical boundaries delineating the appropriate utilisation of morally endowed SBI systems share similarities with the discourse on the ethical treatment of animals. Animals occupy a distinct moral standing; however, their use in research is ethically permitted through regulations that concurrently safeguard their welfare and enable their use in facilitating contributions to scientific advancements that ultimately benefit humanity. This rationale extends congruently to human–animal chimeras, which resemble SBI systems in that they involve the integration of human biological components. We posit that a parallel principle is applicable to SBI systems. Precision in treatment hinges on the salient attributes that underpin their moral standing. If an SBI system exhibits sentience, it should be shielded from suffering, and if it manifests a degree of agency entailing specific needs, these needs should be met. Extensive deliberations within animal ethics have provided insights into pivotal attributes and protective mechanisms (Beauchamp & DeGrazia, 2019) that can serve as a fundamental framework for ethically handling SBI systems, guided by their moral standing.

However, a comprehensive scope of potential SBI technology applications requires a more detailed ethical analysis that is beyond the purview of current animal ethics. For example, given the current trends in AI technologies, military use seems to be one of the most promising applications of SBI. However, we cannot simply rely on existing animal ethics in this area, as the moral permissibility of employing animals in military contexts is not well explored compared to their use in research (Milburn & Van Goozen, 2023). Clearly, intertwined discourses regarding the ethical utilisation of animals, AI systems, and SBI systems require rigorous examinations.

Additionally, the moral evaluation of SBI systems mandates scrutiny of not only their eventual applications but also the training processes. Analogous to the ethical predicament associated with overtraining or constraining animals, training morally endowed SBI systems may raise ethical concerns. As discussed above, it may be possible that SBI systems could be trained such that they experience wrong responses as pain, and correct responses as pleasure. Training SBI systems by exposing them to repeated painful stimuli, is analogous to training animals by hurting them when they get wrong answers. While this is a very speculative concern, it is important to consider speculative future cases in order to set parameters of the responsible development of SBIs.

3. Non moral standing-related ethical dimensions of SBI systems

There are many ethical quandaries faced by SBI systems beyond moral standing, which warrant scrutiny.

3.1 Consent

The indispensable need of SBI technologies for human biological materials distinguishes them from AI technologies and underscores the significance of consent (Kagan et al., 2023; Savulescu, Gyngell, & Sawai, 2022). Consent encompasses multifaceted concerns intrinsic to cell donation, including ensuring donor autonomy, safeguarding personal data, evaluating potential risks and gains, and determining ownership rights. A pertinent issue intrinsic to SBI technologies is the appropriateness of broad consent, which involves the procurement of authorization for unspecified future applications under predefined conditions. The validity of broad consent has been examined in the context of research on human brain organoids developed in vitro, critiquing the obtainment of human biological materials for uses that may elicit objections from a subset of donors, including research on human brain organoids (Greely, 2020). Notably, the mere creation of human brain organoids evokes sensitivities among certain individuals (Haselager et al., 2020), and extending their use to computational contexts would pose similarly complex challenges. Consequently, the use of SBI technologies requires tailored, case-specific informed consent.

The complexities and evolving nature of SBI technologies would make obtaining informed consent for their use challenging. While donors retain the prerogative to give their consent amid substantial uncertainties, sceptics contend that consent predicated upon excessive ambiguity can raise questions about its legitimacy (Greely, 2021). Nonetheless, during the current and foreseeable developmental phases, the complexity and uncertainty surrounding SBI technologies remain relatively tractable, owing to the avenues for clarification embedded in the consent process (Kagan et al., 2023).

3.2 Donor responsibility for the behaviour of SBI systems

The challenge of ascribing responsibility for harmful behaviours is prevalent in the ethical discourse surrounding AI technologies. This challenge will inevitably extend to SBI technologies as they mature and gain practicality. However, is the issue the same for these systems exactly as that for AI? While AI involves potential responsibility of users, researchers, and developers, or possibly the AI system itself, SBI systems may introduce the additional element of cell donors. The production of SBI systems relies on human neural tissues or organoids hinged upon cell donation. However, the causal and moral responsibilities are distinct. The ethical implications and extent of moral responsibility attributed to a cell donor when an SBI system embodying their cells performs detrimentally would pose a pivotal ethical query intrinsic to SBI technologies.

While a donor merely furnishes cells to an SBI system, researchers or developers establish its specific behaviours. In this context, it appears difficult to directly attribute moral responsibility for behaviour of an SBI system to the donor. However, donors are not mere bystanders; during the consent process, they are informed of the systems their cells contribute to and provide consent for their generation. Therefore, the argument that donors cannot reasonably envision the behaviours of SBI systems falters. Donors are also not inadvertently embroiled in the evolution of these systems nor bereft of autonomous involvement. Moreover, as subsequently discussed, the commercialization of SBI technologies might mandate donor acknowledgment or fair profit sharing, imparting an ethical consideration that favours the moral responsibility of a donor for the behaviour of an SBI system. These nuances propel donors towards being more aligned with the researchers and developers of SBI technologies.

Deciphering the intricacies of donor responsibility requires a comprehensive consideration of these nuances, a formidable task that extends beyond our current purview. Nonetheless, its importance is incontestable, raising the question of whether donors should bear responsibility for the behaviours of SBI systems and provide compensation measures, including financial indemnification; these issues should be thoroughly examined to mitigate potential harmful consequences. Furthermore, donors should be clearly and explicitly informed of their responsibility within the consent process, allowing them to exert substantial influence over their decision to donate cells. Simultaneously, it provides an additional reason, that is, the protection of donor interests, for researchers and developers to avert the detrimental behaviours of SBI systems. We propose that ethicists should thoroughly explore issues related to donor responsibility before negative societal effects associated with using SBI technologies materialize.

An ancillary concern is whether donors are perceived by people as morally culpable for the behaviours of SBI systems. This is still a speculative idea; perhaps one might not make such an attribution of responsibility, just as one would not hold parents responsible for the behaviour of their children solely because of a genetic connection. Nonetheless, the issue is worth raising because even if donors are not materially responsible for the behaviour of these systems, the public perception might suggest otherwise, creating ethical dilemmas. Currently, there is clear societal aversion towards specific AI technologies, evidenced by lawsuits against generative AI enterprises and baseless online defamation campaigns targeting AI researchers. In this context, if the idea that cell donors should shoulder the responsibility for the behaviour of SBI systems becomes widespread, it could potentially lead to unjust reprisals against the donors. This underscores the urgency of careful ethical analysis concerning donor responsibility for SBI system conduct coupled with its judicious dissemination, and the stringent protection of donor's personal data.

In the absence of such ethical provisions, the issues related to donor responsibility could be bypassed should researchers or developers utilize their own cells for the creation of neural tissues for integration into SBI systems. As researchers involved in SBI research are best placed to understand the potential implications of donation for this purpose, this could be a morally prudent approach, at least in the interim.

3.3 Commercialization and benefit sharing

Private enterprises and public research institutions have already started developing SBI technologies (Kagan et al., 2023). The prospect of significant monetary value from SBI technologies looms on the horizon. If this were to happen, how should we reward and compensate cell donors?

This predicament echoes a similar discussion in research on human brain organoids, a field that is expected to generate significant financial gains through their medical applications. Donors are often excluded from reaping these rewards due to the absence of recognized property rights pertaining to donated cells. Nonetheless, the projected substantial commercial worth of technologies related to human brain organoids has sparked demands for sharing benefits with donors (Boers et al., 2016).

Similarly, donors contributing cells to SBI technologies may warrant benefit sharing. However, identifying the appropriate forms of such sharing presents a considerable challenge. In the realm of the medical applications of SBI technologies, one approach entails giving donors a higher priority for medical breakthroughs made possible through new research. However, as with current AI technologies, many applications of SBI technologies are likely to be nonmedical. Another option is to allow donors to share in the profits made possible from these technologies. (Kinderlerer, 2023; Savulescu et al., 2022).

Integrating human cells into computer applications presents nuanced legal challenges, particularly concerning the patentability of SBI systems. The issue of patenting stem cell-related materials remains intricate, and the question of whether—and to what extent—human brain organoids can be patented is yet unresolved (Wolff, 2024). The emergence of SBI technologies is likely to amplify these complexities. Disparities in patent decisions across different jurisdictions could introduce fairness concerns. Additionally, the question of whether cell donors ought to be involved in patent applications for specific SBI systems warrant further exploration.

The commercialization of SBI technologies is associated with many ethical considerations, ranging from donor entitlements and extant regulations governing human biological materials to prevailing public sentiment. Given the profound potential benefits accompanying the evolution of SBI technologies, reconciling interests and rights among diverse stakeholders has become an imperative mandate since its inception. This perspective has garnered substantial recognition among researchers, including SBI developers, prompting diverse initiatives and visionary endeavours (Hartung et al., 2023; Kagan et al., 2023; Morales Pantoja et al., 2023; Smirnova et al., 2023).

4. Conclusion

In this chapter, we outline some of the complex ethical challenges raised by SBI technologies. While SBI technologies are in their nascent stage, and the concerns we discuss are speculative, it is crucial to consider a wide range of possible future scenarios with SBI to help inform discussions about the long-term development of this technology.

The unique features of SBI may render them especially vulnerable to concerns involving moral standing. The integration of neuron and silicon components allows for the creation of feedback between external stimuli and internal states, which could plausibly encourage the development of subjective experiences of pleasure and pain distinction.

SBI technologies raise unique ethical challenges in relation to cell donors. Given their multiple potential harmful and beneficial applications, it raises challenging issues about responsibility, patenting, and commercialization, which must be addressed as these technologies are developed.

Empirical studies, looking at the expectations and attitudes of stakeholders, also need to be performed to help influence the responsible development of SBI technologies. The extent to which individuals attribute

Moral dimensions of synthetic biological intelligence

agency to SBI systems, their levels of endorsement of SBI technologies, their willingness to donate cells for using SBI technologies, the attribution of responsibilities for the behaviours of SBI systems to donors, and the concept of fairly sharing benefits within SBI technologies all necessitate further empirical analysis and investigation. The way that the media portray SBI technologies also warrants observation. Fostering diverse avenues for public engagement vis-à-vis SBI technologies is pivotal to ensuring that its evolution manifests in a socially responsible manner. Just as AI perpetually reshapes our societal fabric, the eventual impact of SBI technologies may be unpredictable. It is important to explore the possible implications of SBI systems while we are still in a position in influence their developmental trajectory.

Acknowledgements

TS was funded in whole, or in part, by the Japan Agency for Medical Research and Development (AMED) [Grant Number JP23wm0425021], the Japan Society for the Promotion of Science (JSPS) KAKENHI [Grant Number 21K12908], JST Research Institute of Science and Technology for Society (RISTEX) [Grant Number JPMJRS22J4], the Uehiro Foundation on Ethics and Education [Grant Number UEHIRO2023–0110]. This research is also supported by the Wellcome Trust [Grant Number WT203132/Z/16/Z] and the Singapore Ministry of Health's National Medical Research Council under its Enablers and Infrastructure Support for Clinical Trials-related Activities Funding Initiative (NMRC Project No. MOH-000951–00). MK received funding from Research Center on Ethical, Legal and Social Issues in Osaka University through ELSI Co-Creation Research Fund (Theoretical Study for Constructing Dual-use Ethics of an International Standard: Focusing on the Dialogue between Normative and Descriptive Research). CG and JS, through thier involvement with the Murdoch Children's Research Institute, received funding from the Victorian State Government through the Operational Infrastructure Support (OIS) Program.

Competing interests

JS is a Partner Investigator on an Australian Research Council grant LP190100841 which involves industry partnership from Illumina. He does not personally receive any funds from Illumina. JS is a Bioethics Committee consultant for Bayer. The other authors declare that there is no conflict of interest.

References

Beauchamp, T. L., & DeGrazia, D. (2019). *Principles of animal research ethics*. Oxford: Oxford University Press.

Boers, S. N., van Delden, J. J., Clevers, H., & Bredenoord, A. L. (2016). Organoid biobanking: Identifying the ethics. *EMBO Reports, 17*(7), 938–941. https://doi.org/10.15252/embr.201642613.

Buchanan, M. (2018). Organoids of intelligence. *Nature Physics, 14*, 634. https://doi.org/10.1038/s41567-018-0200-2.

Friston, K. J., Wiese, W., & Hobson, J. A. (2020). Sentience and the origins of consciousness: From Cartesian duality to Markovian monism. *Entropy, 22*(5), 516. https://doi.org/10.3390/e22050516.

Gray, H. M., Gray, K., & Wegner, D. M. (2007). Dimensions of mind perception. *Science (New York, N. Y.), 315*(5812), 619. https://doi.org/10.1126/science.1134475.

Greely, H. T. (2020). The dilemma of human brain surrogates: Scientific opportunities, ethical concerns. In A. D'Aloia, & M. C. Errigo (Eds.). *Neuroscience and law: Complicated crossings and new perspectives* (pp. 371–399) Berlin: Springer. https://doi.org/10.1007/978-3-030-38840-9_18.

Greely, H. T. (2021). Human brain surrogates research: The onrushing ethical dilemma. *American Journal of Bioethics, 21*(1), 34–45. https://doi.org/10.1080/15265161.2020.1845853.

Hartung, T., Smirnova, L., Morales Pantoja, I. E., Akwaboah, A., Alam El Din, D.-M., Berlinicke, C. A., ... Zack, D. J. (2023). The Baltimore declaration toward the exploration of organoid intelligence. *Frontiers in Science, 1*, 1068159. https://doi.org/10.3389/fsci.2023.1068159.

Haselager, D. R., Boers, S. N., Jongsma, K. R., Vinkers, C. H., Broekman, M. L., & Bredenoord, A. L. (2020). Breeding brains? Patients' and laymen's perspectives on cerebral organoids. *Regenerative Medicine, 15*(12), 2351–2360. https://doi.org/10.2217/rme-2020-0108.

Hyun, I., Scharf-Deering, J. C., Sullivan, S., Aach, J. D., Arlotta, P., Baum, M. L., ... Lunshof, J. E. (2022). How collaboration between bioethicists and neuroscientists can advance research. *Nature Neuroscience, 25*(11), 1399–1401. https://doi.org/10.1038/s41593-022-01187-2.

Isomura, T., Kotani, K., & Jimbo, Y. (2015). Cultured cortical neurons can perform blind source separation according to the free-energy principle. *PLoS Computational Biology, 11*(12), e1004643. https://doi.org/10.1371/journal.pcbi.1004643.

Kagan, B. J., Kitchen, A. C., Tran, N. T., Habibollahi, F., Khajehnejad, M., Parker, B. J., ... Friston, K. J. (2022). In vitro neurons learn and exhibit sentience when embodied in a simulated game-world. *Neuron, 110*(23), 3952–3969.e8. https://doi.org/10.1016/j.neuron.2022.09.001.

Kagan, B. J., Gyngell, C., Lysaght, T., Cole, V. M., Sawai, T., & Savulescu, J. (2023). The technology, opportunities, and challenges of synthetic biological intelligence. *Biotechnology Advances, 68*, 108233. https://doi.org/10.1016/j.biotechadv.2023.108233.

Kataoka, M., Gyngell, C., Savulescu, J., & Sawai, T. (2023). The importance of accurate representation of human brain organoid research. *Trends in Biotechnology, 41*(8), 985–987. https://doi.org/10.1016/j.tibtech.2023.02.010.

Kataoka, M., & Sawai, T. (2023). What implications do a consciousness-independent perspective on moral status entail for future brain organoid research? *AJOB Neuroscience, 14*(2), 163–165. https://doi.org/10.1080/21507740.2023.2188285.

Kinderlerer, J. (2023). Organoid intelligence: Society must engage in the ethics (version 1). *Frontiers.* https://doi.org/10.25453/plabs.22317001.v1.

Koplin, J. J., & Savulescu, J. (2019). Moral limits of brain organoid research. *Journal of Law, Medicine & Ethics, 47*(4), 760–767. https://doi.org/10.1177/1073110519897789.

Le Page, M. (2021). Human brain cells in a dish learn to play Pong faster than an AI. *NewScientist.* https://www.newscientist.com/article/2301500.

Leufer, D. (2020). Why we need to bust some myths about AI. *Patterns, 1*(7), 100124. https://doi.org/10.1016/j.patter.2020.100124.

Milburn, J., & Van Goozen, S. (2023). Animals and the ethics of war: A call for an inclusive just-war theory. *International Relations, 37*(3), 423–448. https://doi.org/10.1177/00471178231191297.

Morales Pantoja, I. E., Smirnova, L., Muotri, A. R., Wahlin, K. J., Kahn, J., Boyd, J. L., ... Hartung, T. (2023). First Organoid Intelligence (OI) workshop to form an OI community. *Frontiers in Artificial Intelligence, 6*, 1116870. https://doi.org/10.3389/frai.2023.1116870.

Niikawa, T., Hayashi, Y., Shepherd, J., & Sawai, T. (2022). Human brain organoids and consciousness. *Neuroethics, 15*(1), 5. https://doi.org/10.1007/s12152-022-09483-1.

Salomons, T. V., & Iannetti, G. D. (2022). Fetal pain and its relevance to abortion policy. *Nature Neuroscience, 25*(11), 1396–1398. https://doi.org/10.1038/s41593-022-01188-1.

Savulescu, J., Gyngell, C., & Sawai, T. (2022). Tech firms are making computer chips with human cells: Is it ethical? *Conversation.* May 24. https://theconversation.com/tech-firms-are-making-computer-cells-is-it-ethical-183394.

Shepherd, J. (2023). Non-human moral status: Problems with phenomenal consciousness. *AJOB Neuroscience, 14*(2), 148–157. https://doi.org/10.1080/21507740.2022.2148770.

Sinnott-Armstrong, W., & Conitzer, V. (2021). How much moral status could artificial intelligence ever achieve? In S. Clarke, H. Zohny, & J. Savulescu (Eds.). *Rethinking moral status* (pp. 269–289) Oxford: Oxford University Press. https://doi.org/10.1093/oso/9780192894076.003.0016.

Smirnova, L., Caffo, B. S., Gracias, D. H., Huang, Q., Morales Pantoja, I. E., Tang, B., ... Vogelstein, J. T., ... Hartung, T. (2023). Organoid intelligence (OI): The new frontier in biocomputing and intelligence-in-a-dish. *Frontiers in Science, 1*, 1017235. https://doi.org/10.3389/fsci.2023.1017235.

Strubell, E., Ganesh, A., & McCallum, A. (2019). Energy and policy considerations for deep learning in NLP. *Proceedings of the 57th Annual Meeting of the Association for Computational Linguistics* (pp. 3645–3650). (pp. 3645) Florence, Italy: Association for Computational Linguistics.

Wolff, H. (2024). Patentability of brain organoids derived from iPSC: A legal evaluation with interdisciplinary aspects. *Neuroethics, 17*, 7. https://doi.org/10.1007/s12152-023-09541-2.

Zilio, F. (2023). An ontological approach to the ethical issues of human cerebral organoids [version 1; peer review: 1 approved with reservations]. *Molecular Psychology, 2*, 17. https://doi.org/10.12688/molpsychol.17555.1.

Zilio, F., & Lavazza, A. (2023). Consciousness in a rotor? Science and ethics of potentially conscious human cerebral organoids. *AJOB Neuroscience, 14*(2), 178–196. https://doi.org/10.1080/21507740.2023.2173329.

CHAPTER TWELVE

What the embedded ethics approach brings to AI-enhanced neuroscience

Stuart McLennan*, Theresa Willem, and Amelia Fiske

Munich Embedded Ethics Lab (MEEL), Institute of History and Ethics in Medicine, Department of Clinical Medicine, TUM School of Medicine and Health, Technical University of Munich, Munich, Germany
*Corresponding author. e-mail address: stuart.mclennan@tum.de

Contents

1. Introduction	215
2. Embedded ethics	218
3. Applying embedded ethics to AI-enhanced neuroscience	220
4. Conclusion	222
References	223

Abstract

The intersection of neuroscience and artificial intelligence (AI) promises important advances, but it also raises important ethical challenges, including data privacy, bias, accountability, and the implications of cognitive enhancement. To help develop AI-enhanced neuroscience technologies that benefit and do not harm individuals and society, it is important that ethical and social considerations are integrated into research and development of AI-enhanced neuroscience, and not left as an afterthought. In this chapter, we explore how the embedded ethics approach can play an important role in helping to identify and address the ethical, legal, and social issues arising from the integration of AI technology into neuroscience in a deeply collaborative and interdisciplinary manner across the entire development process. We outline important elements of the approach and use a hypothetical case study to demonstrate how embedded ethics can potentially aid in the development of more ethically and socially responsible AI-enhanced neuroscience technologies.

1. Introduction

The increasing application of artificial intelligence (AI) in the field of neuroscience promises exciting advancements in both fields (Macpherson et al., 2021; Surianarayanan, Lawrence, Chelliah, Prakash, & Hewage, 2023). Neuroscience can utilise AI's computational power to analyse large datasets and

Developments in Neuroethics and Bioethics, Volume 7
ISSN 2589-2959, https://doi.org/10.1016/bs.dnb.2024.02.010
Copyright © 2024 Elsevier Inc. All rights reserved, including those for text and data mining, AI training, and similar technologies.

215

obtain meaningful insights, aiding neuroscientists in interpreting complex neural data, making sense of intricate neural pathways, and accelerating the pace of drug discovery for neurological disorders. Conversely, insights from neuroscience can potentially inform the development of more efficient algorithms. These advances may contribute to broader transformations in healthcare, personalised medicine, and our understanding of consciousness itself (Macpherson et al., 2021). The interaction between neuroscience and AI has already resulted in several promising developments, such as brain-computer interfaces (BCIs) that enable paralysed individuals to control robotic limbs, deep learning algorithms that outperform humans in image recognition tasks, and AI-driven drug discovery models (Kim et al., 2019; Lee et al., 2019; Pun, Ozerov, & Zhavoronkov, 2023).

With advances in these fields, important ethical and social challenges have also emerged (Ienca & Ignatiadis, 2020; Hamzelou, 2023). As the collection and analysis of sensitive personal brain data is essential to AI-driven neuroscience, important risks around privacy and data misuse arise. Many of the patients involved in the development and use of AI-enhanced neuroscience technologies have difficulties in communicating and they are often dependent on others to interpret their gaze, winks, or muscle twitches, which can make maintaining patient privacy more difficult. The neurological condition of patients, combined with the complexity and potential consequences of the technology, also make obtaining valid informed consent very challenging. Participants may not fully understand the implications of sharing their brain data for research, and because of the novelty of many of the technologies in this area, many of the potential implications may be unknown. Furthermore, the inherent opacity of many AI algorithms makes it difficult to explain the reasoning behind their recommendations. The development of AI-enhanced neuroscience technologies also raises concerns about fairness and equity. The data used to train AI algorithms can include biases that can potentially result in discriminatory outcomes, particularly when used in clinical applications like disease diagnosis and treatment recommendations. The ability to access such technologies may also exacerbate existing societal disparities, creating a "cognitive divide" which can create differences between people who have access to and can take advantage of technology from those who do not. Additional research is needed to further understand how a potential cognitive divide may intersect with or exacerbate existing socio-economic divisions. The integration of AI in neuroscience technologies also introduces significant safety concerns, including seizures, bleeding, infections, surgical complications, or changes in personality, memory or emotion (Mudgal, Sharma, Chaturvedi, & Sharma, 2020). Malfunctions or errors in AI systems could also result in varying degrees of harm to patients. For example, misinterpretation of brain signals could lead to an increase in errors. In the context of brain signal-to-text outputs, this may simply frustrate users. However, if the outputs are used to trigger automated care tasks, physical harm could occur. When things do go wrong, determining responsibility and accountability poses significant ethical challenges. Furthermore, in cases where devices are implanted in a patient's brain as part of a research trial, in some cases the implants must be removed at the closure of the trial, or additional support must be secured to continue caring for the individual with the implant. This requires medical and financial resources and a commitment on the part of clinicians to continue to care for and work with the patient following the trial. Finally, AI technology developed for neuroscience can have dual-use applications, with the potential to be both beneficial and harmful. For example, some have raised concerns that an AI-enhanced neurological implant that helps paralysed patients speak again could potentially be maliciously used, such as to interpret a person's brain signal against their will (Belluck, 2023).

A key limitation of traditional ethical evaluations of technologies is their retrospective nature. Ethical concerns can only be effectively addressed once they are identified. Unfortunately, the identification of these issues often relies on reacting to the harm they generate, resulting in a chronic delay in providing ethical guidance. However, by the time these issues come to light, it may already be too late, as technologies and their applications may have already become deeply integrated into society, potentially causing real-world harm (Collingridge, 1980). Consequently, ethical analysis is often criticised for lacking proactive strategies to prevent damage before it occurs (Tigard et al., 2023).

In the context of AI-enhanced neuroscience the significant and multifaceted risks raised by this technology have the potential to impact individuals and society in profound ways. The consideration of the ethical issues arising from the integration of neuroscience and AI, therefore, cannot be an afterthought but should be proactively incorporated into the development process from the very beginning. Indeed, the importance of proactively evaluating ethical issues in the context of neuroengineering has already been extensively discussed, and a "proactive ethical design" framework has been proposed emphasising the minimisation of power imbalances, compliance with biomedical ethics, translationality, and social awareness (Ienca, Kressig, Jotterand, & Elger, 2017). In this chapter, we suggest that the "embedded ethics" approach could be beneficial in tackling the ethics concerns raised by AI-enhanced neuroscience in a prospective manner and could help implement frameworks like proactive ethical design (Tigard et al., 2023; McLennan et al., 2022; McLennan et al., 2020; Fiske et al., 2020).

2. Embedded ethics

Embedded ethics refers to the practice of integrating ethics into the entire development process by having ethicists and developers working together to identify and address ethical issues via an iterative and continuous process from the outset of development (McLennan et al., 2022; McLennan et al., 2020). The primary goal of embedded ethics is to facilitate the development of health technologies and interventions that are ethically and socially responsible. To achieve this goal, embedded ethics seeks to guide the design, development, and implementation of a technology or intervention by ensuring that ethical issues are anticipated, assessed, and addressed throughout the project lifecycle (McLennan et al., 2022; McLennan et al., 2020). As part of the Munich Embedded Ethics Lab, we draw from our experience integrating ethics in seven different projects with various focuses, including treating mental dysfunction, developing diagnostic tools for clinical dermatology and radiology, developing polygenic risk scores, identifying biomarkers for chronic skin diseases, developing an algorithm for assisting with moral decision-making, developing intelligent care robotics for older adults, and critically examining AI's responsible design and use in medical imaging. An in-depth description of one of the projects is also provided in this volume's chapter by Franziska Schönweitz, Anja Ruess and Ruth Müller.

Embedded ethics can be implemented in different ways, but the integration should involve regular communication from the early stages rather than seeking ethics input only when problems arise. To gain insight into project teams' structures and workflows, we have found it helpful to actively participate in their regular meetings and, whenever possible, work alongside them in the same physical space. This ongoing engagement with the research team allows embedded ethicists to create opportunities to proactively identify potential ethical issues that may arise during the research and to seek clarifications on technical matters that could impact ethical considerations. Spending time within the research teams' environments also allows embedded ethicists to become acquainted with their colleagues' work culture and engage with them on a social level, fostering mutual trust, shared understanding, and a collaborative working style that forms the basis for addressing ethical questions.

In the context of embedded ethics projects, we have found a range of methods to be particularly helpful in ensuring that ethical considerations are integrated into technology development (Willem et al., 2023). In the initial phase of a project, stakeholder analysis and literature reviews have proven useful in delineating the embedded ethical activities for the course of the project. Stakeholder analyses involve identifying and assessing the various stakeholders involved in a project, such as researchers, patients, companies, and ethics committees. This analysis helps reveal potential dependencies and conflicts of interest among stakeholders. It serves as a foundation for understanding converging and diverging interests in the project and helps focus the concurrent or subsequent literature reviews. These reviews, in turn, are essential for shaping the focus of projects, identifying existing ethical concerns discussed in the literature, gaps in the current understanding of ethical issues, and guiding the focus of empirical research. In the main phase of the project, established empirical ethics and social science methods for data collection and analysis are applied to the research questions previously identified. We have successfully worked with ethnographic approaches, such as participant observation, to gain a deeper understanding of the social and ethical effects of technology development. These methods involve closely observing and documenting social situations related to the project, allowing researchers to gain insights into the perspectives and experiences of stakeholders. Reflexive peer-to-peer interviews helped to facilitate conversations between embedded ethicists and project partners. These interviews create an intimate atmosphere for reflection on the societal relevance and responsibilities associated with the project. However, we found that with this method, challenges such as confidentiality and information gaps may need to be addressed. We conducted focus groups to gather public perceptions and opinions about emerging technologies. They involve moderated group discussions where participants can build on each other's ideas. Scenarios related to the technology are often presented to participants to stimulate discussion and gain insights into their perspectives. Interviews with affected groups and external stakeholders have also shown utility in providing a practice-oriented view and experience. These interviews help contextualise normative analysis, verify literature review findings, and ensure that recommendations are both scientifically robust and relevant to policy and practice. Finally, working directly with the technologies developed in the project is crucial to get a thorough understanding of underlying issues. We, hence, conduct bias analyses to identify and address biases in datasets and algorithms. This includes assessing data representativeness, model performance for under-represented groups, and the potential for omitted-variable bias. The results inform research decisions and can contribute to raising awareness of bias among the participants.

3. Applying embedded ethics to AI-enhanced neuroscience

To illustrate the potential application of embedded ethics in AI-enhanced neuroscience, let's consider a hypothetical situation regarding the restoration of speech following a cataclysmic stroke decades prior. A healthy woman in her 30s suffered a major stroke, which left her paralysed and unable to speak. Two decades later, scientists were able to successfully implant electrodes in her brain, which decode brain signals as she silently says sentences to herself in her head. The technology converts the signals into verbalised speech and changes in facial expressions on an avatar shown on a computer screen. The patient reported that it made her feel "like a whole person again". We have based this scenario on the reporting of a similar case covered in the New York Times, which is based on research looking at the development of technology for strokes or conditions like cerebral palsy and amyotrophic lateral sclerosis (Belluck, 2023; Metzger et al., 2023).

Building on this and similar advances in AI-enhanced neuroscience and large language models, such as attention-based transformer models, our new, hypothetical research project could work towards refining the brain signal-to-text model, which at present is still inaccurate nearly half the time the patient thinks a word. Applying the principles of embedded ethics to the continued development of this technology could involve several steps. Embedded ethics is based on the principle of early integration. In this case, ideally, ethicists, neuroscientists, and AI developers would collaborate from the project's inception to identify ethical issues or particular concerns regarding the privacy of the patients whose data is being used to (pre-)train the model, ability to obtain informed consent from the patients the model is being tested with, the safety of the devices, the accuracy of the decoded speech, potential risks associated with use including misrecognition, false results, changes in personhood, additional financial or emotional costs for the patient, and processes surrounding data security.

The embedded ethicist can also assist researchers in promoting transparency and accountability. Here, processes are identified and implemented through which the project team maintains transparency by documenting the sources of data and the algorithms used. Any biases in the data or model are acknowledged and addressed in the system's design. The embedded ethicist could assist the developers in conducting exercises to increase the transparency of the dataset and algorithms involved in the technology, such as the "datasheets for datasets" and "model cards for model reporting" (Mitchell et al., 2019; Gebru et al., 2021). For the creation of such documents, the embedded ethicist could facilitate a workshop for the team. The datasheets and/or model cards, could then be included as appendices to the published research papers of the project and facilitate reflections among team members about their engagement in the transparency exercises.

One of the benefits of the embedded ethics approach is the accompaniment of the ethicist or social scientist throughout the lifetime of the project. By engaging in continuous monitoring, scientists are responsible for maintaining the ethical standards of the project, which have been outlined and discussed in detail with the embedded ethicist. The embedded ethicist is particularly helpful in keeping the focus on issues identified as being particularly difficult or concerning in relation to neuroscience and the use of AI. Any unexpected consequences, such as unintended cognitive enhancements, are identified and addressed promptly.

AI has the possibility for significant personal and societal impact. To best assess crucial ethical questions from a range of perspectives, embedded ethics utilises stakeholder engagement, which has already been extensively discussed in the context of neuroethics (Boyd & Sugarman, 2022; Das et al., 2022; Giordano & Shook, 2015; Morein-Zamir & Sahakian, 2010). The embedded ethicist also would have the opportunity to engage with patients, caregivers, and rehabilitation professionals to understand their needs and concerns. This could take the form of one-on-one interviews, participant observation, or focus groups, creating opportunities whenever possible for engaged stakeholders to help ensure that neuroscience endeavours are designed for the benefit of society (Das et al., 2022). Importantly, the design of these interactions should build upon attributes such as humility, openness, reflexivity, and others that were demonstrated to support neuroethics engagement specifically (Das et al., 2022), and incorporate ideas from established ways of engaged research, such as the Good Participatory Practices framework (Boyd & Sugarman, 2022). The embedded ethicist then brings these obtained insights to the ongoing ethical analysis of the project as it develops, helping the project team to make changes, address particular concerns, and implement ethical guidelines.

By iteratively integrating empirical research on ethical and social concerns as they emerge within this research study, the embedded ethicist can also help to provide ongoing education and training for those on the development side. The embedded ethicist can provide targeted training to the neuroscientists and AI developers in ethical decision-making to ensure they are well-equipped to address ethical dilemmas as they arise in the current project and future research and clinical applications.

By applying embedded ethics to this hypothetical project, the research team can proactively address ethical issues and create an AI system that not only enhances patient outcomes but also respects individual autonomy, privacy, and the broader societal implications of cognitive rehabilitation. Drawing on insights gained from specific projects such as the one described above and accumulated professional experience, embedded ethicists can also actively contribute to ongoing conversations on the ethics of AI and neuroscience, not only within the research community but also at the level of policymakers and regulatory bodies, and potentially specific publics such as patient groups.

4. Conclusion

The application of AI in neuroscience holds great potential for advancing our understanding of the brain, as well as potentially improving the quality of life for individuals affected by neurological disorders. However, it is already clear that the ethical issues that emerge in this research are far too complex and significant to be addressed as an afterthought. We need to proactively identify and address ethical concerns within the research development process itself, echoing previous calls for proactive ethical design in neuroengineering (Ienca et al., 2017). Embedded ethics offers a promising methodology to help proactively navigate the ethical terrain of AI-driven neuroscience. Integrating ethical considerations into the research and development process can help promote the benefits while mitigating potential harms.

While the precise form of any embedded ethics work depends on the specific constellation of stakeholders, the nature of the research or application under development, and the socio-political situation within which the embedded ethicist is working, several key aspects of embedded ethics should remain constant. The first is that embedded ethics seeks to promote technologies that are ethical-by-design, such that ethical considerations are integrated into the initial stages of project planning and development. This includes defining ethical goals and constraints, as well as considering the potential societal impacts of the technology. Secondly, embedded ethics takes stakeholder engagement seriously, involving stakeholders from diverse backgrounds, including ethicists, policymakers, and affected communities, in the decision-making process. This ensures that a broad spectrum of perspectives is considered when shaping the development and use of AI technology in neuroscience. Thirdly, embedded ethics is a process of continuous assessment and adaptation, which recognises that ethical considerations are not static and will likely change as AI technologies continue to advance. It advocates for ongoing assessment and adaptation of ethical guidelines as technology evolves and as our understanding of its ethical implications deepens and is attuned to the identification of potentially emergent ethical concerns that are not yet fully recognised. Fourthly, embedded ethics promotes transparency in decision-making processes and accountability for the ethical outcomes of AI applications in neuroscience. This includes mechanisms for auditing, reporting, and correcting ethical lapses. Finally, embedded ethics is not just about ethical evaluation, but it also aims to foster enhanced ethical reflection by all project members through education and training. To implement embedded ethics effectively, it is crucial to educate and train researchers, developers, and users about ethical principles and best practices in AI-driven neuroscience to foster a culture of ethical responsibility.

References

Belluck, P. (2023, August 23). A stroke stole her ability to speak at 30. A.I. Is helping to restore it years later. *The New York Times.* ⟨https://www.nytimes.com/2023/08/23/health/ai-stroke-speech-neuroscience.html⟩.

Boyd, J. L., & Sugarman, J. (2022). Toward responsible public engagement in neuroethics. *AJOB Neuroscience, 13*(2), 103–106.

Collingridge, D. (1980). *The social control of technology.* London: Pinter.

Das, J., Forlini, C., Porcello, D. M., Rommelfanger, K. S., Salles, A., Delegates, G. N. S., & Yin, J. (2022). Neuroscience is ready for neuroethics engagement. *Frontiers in Communication, 7,* 909964.

Fiske, A., Tigard, D., Müller, R., Haddadin, S., Buyx, A., & McLennan, S. (2020). Embedded ethics could help implement the pipeline model framework for machine learning healthcare applications. *American Journal of Bioethics, 20*(11), 32–35.

Gebru, T., Morgenstern, J., Vecchione, B., Wortman Vaughan, J., Wallach, H., Daumé, I. I. I., H., & Crawford, K. (2021). Datasheets for datasets. *Communications of the ACM, 64*(12), 86–92.

Giordano, J., & Shook, J. R. (2015). Minding brain science in medicine: On the need for neuroethical engagement for guidance of neuroscience in clinical contexts. *Ethics in Biology, Engineering and Medicine: An International Journal, 6*(1–2), 37–41.

Hamzelou, J. (2023, May 25). *A brain implant changed her life. Then it was removed against her will: Her case highlights why we need to enshrine neuro rights in law.* MIT Technology Review. ⟨https://www.technologyreview.com/2023/05/25/1073634/brain-implant-removed-against-her-will/⟩.

Kim, M., Yun, J., Cho, Y., Shin, K., Jang, R., Bae, H. J., & Kim, N. (2019). Deep Learning in Medical Imaging. Neurospine. *Neurospine. 16*(4), 657–668. https://doi.org/10.14245/ns.1938396.198 Epub 2019 Dec 31. Erratum in: Neurospine. 2020 Jun;17(2):471-472. PMID: 31905454; PMCID: PMC6945006.

Ienca, M., & Ignatiadis, K. (2020). Artificial intelligence in clinical neuroscience: Methodological and ethical challenges. *AJOB Neuroscience, 11*(2), 77–87.

Ienca, M., Kressig, R. W., Jotterand, F., & Elger, B. (2017). Proactive ethical design for neuroengineering, assistive and rehabilitation technologies: The Cybathlon lesson. *Journal of Neuroengineering and Rehabilitation, 14*(1), 115.

Lee, M. B., Kramer, D. R., Peng, T., Barbaro, M. F., Liu, C. Y., Kellis, S., & Lee, B. (2019). Brain-computer interfaces in quadriplegic patients. *Neurosurgery Clinics of North America, 30*(2), 275–281.

Macpherson, T., Churchland, A., Sejnowski, T., DiCarlo, J., Kamitani, Y., Takahashi, H., & Hikida, T. (2021). Natural and artificial intelligence: A brief introduction to the interplay between AI and neuroscience research. *Neural Networks, 144*, 603–613.

McLennan, S., Fiske, A., Tigard, D., Müller, R., Haddadin, S., & Buyx, A. (2022). Embedded ethics: A proposal for integrating ethics into the development of medical AI. *BMC Medical Ethics, 23*, 6.

McLennan, S., Fiske, A., Celi, L. A., Müller, R., Harder, J., Ritt, K., ... Buyx, A. (2020). An embedded ethics approach for AI development. *Nature Machine Intelligence, 2*, 488–490.

Metzger, S. L., Littlejohn, K. T., Silva, A. B., Moses, D. A., Seaton, M. P., Wang, R., ... Chang, E. F. (2023). A high-performance neuroprosthesis for speech decoding and avatar control. *Nature, 620*(7976), 1037–1046.

Mitchell, M., Wu, S., Zaldivar, A., Barnes, P., Vasserman, L., Hutchinson, B., ... Gebru, T. (2019). Model cards for model reporting. *Proceedings of the Conference on Fairness, Accountability, and Transparency (FAT* '19), USA*, 220–229.

Morein-Zamir, S., & Sahakian, B. J. (2010). Neuroethics and public engagement training needed for neuroscientists. *Trends in Cognitive Sciences, 14*(2), 49–51.

Mudgal, S. K., Sharma, S. K., Chaturvedi, J., & Sharma, A. (2020). Brain computer interface advancement in neurosciences: Applications and issues. *Interdisciplinary Neurosurgery, 20*, 100694.

Pun, F. W., Ozerov, I. V., & Zhavoronkov, A. (2023). AI-powered therapeutic target discovery. *Trends in Pharmacological Sciences, 44*(9), 561–572.

Surianarayanan, C., Lawrence, J. J., Chelliah, P. R., Prakash, E., & Hewage, C. (2023). Convergence of artificial intelligence and neuroscience towards the diagnosis of neurological disorders—A scoping review. *Sensors, 23*(6), 3062.

Tigard, D., Braun, M., Breuer, S., Ritt, K., Fiske, A., McLennan, S., & Buyx, A. (2023). Toward best practices in embedded ethics: Suggestions for interdisciplinary technology development. *Robotics and Autonomous Systems, 167*, 104467.

Willem, T., Fritzsche, M. C., Zimmermann, B. M., Sierawska, A., Breuer, S., Braun, M., & Buyx, A. (2023). Embedded ethics in practice: A toolbox for integrating the analysis of ethical and social issues into healthcare AI research. Unpublished manuscript.

> CHAPTER THIRTEEN

From being embedded in practice: Working situated and interdisciplinary in the neurosciences and neurocomputation as ethicists and social scientists

Franziska B. Schönweitz[a,*], Anja K. Ruess[b,c], and Ruth Müller[b,c]

[a]Institute of History and Ethics in Medicine, TUM School of Medicine and Health, TUM School of Social Sciences and Technology, Technical University of Munich, Munich, Germany
[b]Department of Science, Technology and Society (STS), TUM School of Social Sciences and Technology, Technical University of Munich, Munich, Germany
[c]Department of Economics and Policy, TUM School of Management, Technical University of Munich, Munich, Germany
*Corresponding author. e-mail address: franziska.schoenweitz@tum.de

Contents

1. Introduction	226
2. Situated knowledges and interdisciplinary research	228
3. Embedded ethics and social sciences in practice—active integration in frontier neuroscience and neurocomputation	230
4. Investigating neuroethics and ethics of AI within NEUROTECH: Data perception, security, and representativeness	232
5. The need for situatedness of ethics and social science research in complex network structures	236
6. Conclusion	238
Acknowledgments	238
References	238

Abstract

Engaging with ethical and social questions is becoming increasingly important, especially in health research. The interdisciplinary context of this field unites computer scientists, engineers, physicians, biochemists, or neuroscientists. This, in turn, has implications on how individual types of knowledge emerge within project work and how topics are situated in a societal and ethical sense. Here, we draw on a specific project to highlight the difficulties of interdisciplinary work and how they can be addressed in the context of embedded ethics. The Technical University of Munich's Innovation Network for Neurotechnology in Mental Health (NEUROTECH) is one of the

Developments in Neuroethics and Bioethics, Volume 7
ISSN 2589-2959, https://doi.org/10.1016/bs.dnb.2024.02.001
Copyright © 2024 Elsevier Inc. All rights reserved, including those for text and data mining, AI training, and similar technologies.

university's recent projects promoting the integration of embedded ethics and social sciences. The network operates at the intersection of AI and neuroscience and is concerned both with the application of invasive and non-invasive neurotechnologies for the study of mental disorders and with the development of Machine learning algorithms and mathematical models based on the data obtained by the network. Through our active participation and fieldwork as embedded ethicists and social science researchers, we find that despite similarities in overarching themes, for instance data perception, collection, security, and privacy, the mathematical, computer science, and neurological research groups tend to conceptualise these themes differently, depending on their often-multiple professional socialisations. Based on our observations, we advocate for a situated perspective on emerging ethical and social issues to meet the individual needs of each context.

1. Introduction

It is a sunny morning in September, when we get off the bus at the campus of the University of Bonn, gazing at the modern glass building of the Deutsches Zentrum für Neurodegenerative Erkrankungen (German Center for Neurodegenerative Diseases), where the Nature Conference "AI, Neuroscience and Hardware" would take place. A bit we feel like outsiders, attending the conference as embedded ethicists and social science researchers with neither AI nor neuroscience being our primary scientific background. This feeling should change, though, by the time of the first coffee break when we establish contacts and thereby become confronted with no small expectations towards our role: [...] "We speak the same language in neuroscience and AI most of the time, I would say. But ethics (.) don't you have some sort of overarching perspective, too? Aren't you perhaps the glue that brings the whole thing together?" (DZNE, Field Note, 07.09.2022).

As illustrated by this fieldnote from the 2022 Nature Conference *AI, Neuroscience and Hardware,* neuroscience and AI are often portrayed as related research areas that "have closely interacted and inspired each other since the beginning of AI" (DZNE, 2022). At the heart of this relationship is the notion and metaphor often invoked in neuroscience research: that a computer can be thought of as a brain and, respectively, a brain can be thought of as a computer. Both disciplines' deeply interwoven trajectories are expressed in research goals such as augmenting human intelligence with machines that are capable of learning, as well as computing techniques that explicitly relate to the functioning of the human brain. Concurrently, we are witnessing an upswing of both neuro-inspired AI technologies (e.g. neural networks) and neuro-technologies with AI components (e.g. brain–computer interfaces) that increasingly permeate various areas of social life. Medicine provides an important field of developing and applying these technologies, with numerous potential use cases. For example, scientists have begun to use neurotechnologies to restore degenerated motor functions, decode speech, or diagnose and treat mental dysfunction, such as chronic pain (Lorach, Galvez, & Spagnolo, 2023; Metzger, Littlejohn, & Silva, 2023; Yoshida, Hashimoto, & Shikota, 2016). The notion of AI that we use in the following is very broad, ranging in architecture from knowledge-based, that is stochastic and logical, methods to trainable Machine Learning (ML) algorithms and neural networks.

While the two disciplines, neuroscience, and AI, are not only deeply intertwined but often thought of together, our field note also points to another aspect of this relationship: the role of ethical inquiry. The scientist cited above, who had a background in ML, conceived of ethics not only as a relevant issue in both neuroscience and AI but also as an overarching one that applies to both fields in similar ways, for example by raising similar questions and concerns for both fields. Echoing our interlocutor, the scientific literature provides similar accounts that call for an integrated and cross-disciplinary ethical approach to both neuroscience and AI based on their alleged similarities and interdependencies (Farisco et al., 2022; Weh & Soetebeer, 2023). Other authors, particularly from the fields of ethics and social science, take a more cautious stance: While they consider the nexus of the two disciplines to be useful for developing and discussing ethical issues, they emphasise divergent ethical considerations across the two fields (Berger & Rossi, 2023). From this perspective, the question arises as to how ethical issues that emerge in and across the two disciplines relate to each other and how they pertain to the work of ethicists and social science researchers at the interface of neuroscience and AI.

Although the focus on interdisciplinary work in healthcare holds great promise, proponents of this approach recognise that there will be inevitable challenges, such as ethical dilemmas (Smye & Frangi, 2021). Depending on the research perspective, different ethical challenges or even divergent assessments arise. Finding a consensus can thus be a significant task. Further challenges are concerned with different research practices, especially when juxtaposing sciences and humanities, but also when it comes to the different science fields including the "[...] numerous structural, cultural and organisational features of the university environment" (Hall et al., 2006, p. 767). Researchers in interdisciplinary collaborations commonly bring together a variety of different ontologies, epistemologies, logics and internalised practices (Barry, Born, & Weszkalnys, 2008). They tend to have different understandings and visions of the purpose and the appropriate methods of interdisciplinary collaboration (Callard et al., 2015). Not least, this is reflected in questions about where and how to public research findings emerging from interdisciplinary collaborations (Callard & Fitzgerald, 2015).

To tackle these issues, Embedded Ethics and Social Sciences, in which the authors of this contribution were and are involved in developing, is gaining ground in technology development. This is the case for academia and the industry (Fiske et al., 2020; McLennan et al., 2020, 2022; Mökander & Floridi, 2023; Tigard et al., 2023). It implies the active involvement of ethicists and social science researchers to tackle emerging questions and avoid ethical afterthought. This approach is particularly visible in healthcare (Fiske et al., 2020; McLennan et al., 2022; Tigard et al., 2023), working with new frontier technologies in neuroscience and related computational models, including AI systems, to tackle a wide variety of diseases and health restrictions (Berberich, Paredes-Acuna, Lipp, & Cheng, 2023; Ploner et al., 2023). Drawing on Embedded Ethics and Social Science research allows us to observe how ethical questions arise in situated contexts and through situated practices.

In this paper, we draw on an interdisciplinary network project, which we regard as a specific apparatus of knowledge production. Our chapter aims to highlight the challenges of situated knowledges and interdisciplinarity by observing how ethical questions emerge from the highly situated interactions within a particular research project and how they relate to the epistemologies represented by different researchers. We begin our chapter by providing some theoretical background on situated knowledges and interdisciplinarity. We then summarise our current practical efforts and active integration of embedded ethics and social sciences, before turning to our thematic example, highlighting the parallels and differences in how the network applies the ethics of AI and neuroethics. In doing so, we emphasise how complex and entangled emerging subthemes often are despite their overarching similarities. Building on our analysis, we discuss the varying conceptions of ethics by pointing to the current state of research on situatedness and situated ethics. We conclude with what doing embedded ethics and social sciences can and should mean in a cross-disciplinary setting.

2. Situated knowledges and interdisciplinary research

To understand how situated contexts shape emerging ethical considerations in interdisciplinary research, we consider Donna Haraway's notion of *situated knowledges* particularly helpful. Focusing on the late 1980s scientific community in the United States, Haraway first coined the term, challenging the then-widespread ideal of epistemic neutrality and scientific objectivity that transcends contexts. Instead, she proposed a different angle on knowledge production that considers the context, for example historical, cultural, disciplinary, in which the knowing person is embedded or, put differently, situated (Haraway, 1988). Our position within a specific context shapes our knowledge and understanding of truth, thus rendering the formation of knowledge inherently positional and embodied. Rather than rejecting objectivity altogether, Haraway proposes a feminist understanding of objectivity that stems from the doctrine of the partial, situated perspective of the knowing person: "only partial perspective promises objective vision" (Haraway, 1988, p. 583) that is both locatable and accountable.

More concretely, in the context of interdisciplinary research endeavours, a person creates a unique understanding of their knowledge and how their expertise intersects with other fields or collaborators. As we will explicate in this chapter based on our own empirical work, these situated viewpoints of researchers tend to be interrelated with their individual scientific, often disciplinary socialisations. To make sense of these viewpoints, we find Merton's notion of scientific *ethos* (Merton, 1973) particularly helpful. Robert Merton uses the concept of scientific ethos to describe the non-codified norms and values internalised by scientists and practiced in their daily work routines (Merton, 1973). These norms, values and practices shape the scientists' situated ways of knowing and producing knowledge. At the same time, when considering interdisciplinary research, the process of knowledge production is not a separate one of individuals but takes place within a huge apparatus involving unique combinations of researchers, their socialisations, materials, and infrastructures (Barad, 1999). Thus, researchers not only learn from within their ethos and their disciplines but are further influenced by collaborative work, overcoming the boundaries of their initial research focus.

In the following, we will turn to a project highlighting the challenges of situated knowledges and interdisciplinary research, namely the NEURO-TECH Innovation Network. By providing preliminary insights into our ongoing work, we show how ethical and societal issues are conceptualised across neuroscience and neurocomputation research and what *doing* situated, embedded ethics and social science research within an interdisciplinary project may mean in practice.

3. Embedded ethics and social sciences in practice—active integration in frontier neuroscience and neurocomputation

One example of an initiative that aims to embed ethics and Social Sciences into the research process is the NEUROTECH Innovation Network, launched by the Technical University of Munich in 2022. The network focuses on risk-related, explorative neurological research questions that could impact society, covering conditions from chronic pain and depression to aphasia and speech impairments after stroke (Ploner et al., 2023). It consists of two central research groups in neurophysiology with direct patient contact, one involving invasive neurotechnologies and the other using non-invasive methods,[1] and two affiliated groups working on neurocomputation, that is mathematical modelling and ML development, and neuroethics respectively. Within the two main research groups, there is a range of different expertises, including medicine, surgery, psychology, neuroelectronics, neurophysiology, linguistics, and neurocomputation. The two other affiliated groups represent neuroscience and biochemistry backgrounds and expertise in programming, mathematical modelling, social sciences, and ethics. Within all groups, single individuals often have multiple professional socialisations and expertises, for instance a neurophysiology researcher who is also a medical doctor.

The non-invasive neurophysiology group uses electroencephalography (EEG) caps to research chronic pain and its effects on the brain, as well as potentially related mental disorders such as depression. Their research involves approximately 100 chronic pain patients and healthy subjects. In the affiliated group on neurocomputation, the data gathered is used to develop an ML algorithm to detect biomarkers in the EEG recordings of patients with depression or chronic back pain to potentially cluster and stratify them according to their clinical presentation. In contrast, the invasive neurophysiological team uses an intracranial implant to study the language compensation of aphasic patients[2] in their healthy right brain hemisphere. The invasive team collaborates with another part of the neurocomputation group, which is building mathematical mechanistic models of (disordered) language

---

[1] We are aware that invasiveness and non-invasiveness are actors' categories, with the question of the definition of "invasiveness" not satisfactorily resolved in neuroethics (Collins & Klein, 2023; Davis & Koningsbruggen, 2013) and the notion and boundary of what counts as invasive not necessarily shared by all actors involved. In this chapter, we will nonetheless stick with the actors' categories for simplification and, in the following, refer to the two central research groups as the non-invasive and invasive teams.

[2] The research group is currently working with one aphasic patient but plans to enroll a second one.

functions. These models include variables and parameters that cannot be measured by the neuronal recordings, meaning they are aimed at advancing the researcher's understanding of the cognitive processes beyond what can be captured by experimental techniques. Additional network-external collaborations aim to develop AI algorithms that can predict which words or phrases the patient intends to use but cannot formulate due to their clinical condition. This line of research could eventually enable patients with aphasia to have more fluent conversations by assisting them to find and formulate language.

Our approach within the project builds on interdisciplinary upstream integration of ethics and social science researchers into the basic science, clinical, and technology development of research processes (McLennan et al., 2020; McLennan et al., 2022). Methodologically, it includes lab observations and visits of regular meetings of the involved research teams, as well as qualitative semi-structured interviews with researchers from different career stages. We also observe measurements and data collection with patients or healthy subjects, as well as interviews with them. We complement this research by focus groups with patients, healthy subjects, and members of the broader public to gain insight into their perceptions of neuroscience research and neurotechnological development. Moreover, our research includes regular cross-disciplinary literature reviews and exchanges with the scientific communities, for example by attending conferences or publishing peer-reviewed scientific articles, often in co-authorship with our neuroscience and neurocomputation colleagues.

Working in embedded ways also entails sharing our findings and (ad hoc) feedback with the involved research teams. This is to help identify any ethical and social issues as they arise, reflect collectively about how to tackle them, and ultimately, address them throughout ongoing research and development processes. Examples of such feedback have so far included questions of racial and ethnic bias in neuroscience research (Müller et al., 2023). The issue was raised by the non-invasive team, which collects larger dimensions of data and is therefore concerned about their representativeness. Another exemplary issue that arose with the invasive team is the question of what implications the (early) end of the clinical trial might have for their patients (Gilbert, Tubig, & Harris, 2022; Gilbert, Ienca, & Cook,; 2023).

Based on our work as embedded ethicists and social scientists within NEUROTECH, we aim to trace how ethics is conceptualised in and across the initiative, focusing on the similarities and differences between the neurocomputation and neurophysiological research groups.

4. Investigating neuroethics and ethics of AI within NEUROTECH: Data perception, security, and representativeness

Within the network, and despite their overlapping expertises, neurophysiologists and neurocomputation scientists assume different positions in data collection, dissemination, and the processing pipeline. While neurophysiologists work directly with patients and healthy subjects and with the neurotechnologies involved, neurocomputation scientists become involved towards the end of the data processing pipeline. They use data that has already been collected and pre-processed by the invasive or the non-invasive team for further analysis, for example by applying an algorithmic architecture of ML systems, or by developing mathematical models. Although both teams' work is based on the same patient data, our research suggests that researchers hold different conceptualisationss of study participants, whether patients or healthy subjects, and that neurophysiologists exhibit more intimate relationships with participants involved in their research. Especially in the invasive neurophysiology team, there seems to be a strong personal bond between them and their patient. The following quote from a conversation with the responsible PI illustrates this strong relationship:

> *And that's exactly the same with [the patient], that's [the patient]'s original motivation [...]. Not because [the patient] can speak [again], but because [the patient] realizes that they can work at the frontiers of knowledge and participate and is part of the team, [the patient] is our most important employee. Without [the patient], nothing works. (PI02)*

What is particularly striking about this quote is that the PI refers to the patient not only as an essential part of the research but also refers to them as the team's "most important employee". This highlights a conceptualisation of the patient's role differing from traditionally dyadic constructions of researcher-patient relationships (Parsons, 1951) and instead suggests a certain level of parity between researchers and research subjects in the research setting. The PI directly links the patient's role to the ambition to advance the state of research in neuroscience and constructs this as the primary motivation of the patient herself, creating a deliberate distinction from another possible motivation for the patient's participation in the study, namely an improvement in their health status. Still, the PI also emphasises that a partial recovery is not ruled out and that the patient has the best prerequisites for such improvement. In particular, the ways and frequency in which the question of health recovery is addressed among neurophysiologists show that, in addition to the dedicated research objectives of the study in which the patient is embedded out of purely scientific interest, there is also a subtle desire to care for the patient in a medical sense. Thus, we can not only see a general closeness to the patient through the position of the neurophysiologists in the data pipeline but also multiple professional socialisations as both researchers and physicians, whereby particularly the latter conditions a caring relationship with the patient.

This conceptualisation of the patient's role contrasts with the perspectives of the neurocomputation scientists we spoke to. Rather than the medical definitions of disease or patients' clinical pictures and lived realities, neurocomputation scientists tend to foreground the parameters needed to determine their diagnosis. Consider the following quote from an interview with one of the responsible PIs:

> And I have to say quite honestly that we are very, very far away from the real definition [of mental health], because we see this whole problem as a kind of classification problem. That means, for example, if you ask any student, they have an Excel sheet that says zero or one. (PI03)

Particularly, the formulation "an Excel sheet that says zero or one" illustrates how distant neurocomputation scientists position themselves from patients. For them, patients remain primarily data points, while their realities and experiences play no or only a minor role in the network's neurocomputation groups. However, solely basing this conception on the positioning of neurocomputation scientists in the data pipeline falls short. Rather, the ways of working as and being a computation scientist, which have been learned and rehearsed throughout their careers and which simply do not involve direct interactions with patients, differ from those of neurophysiologists. Importantly, this does not mean the neurocomputation scientists lack sense for the lived realities of patients. Several of the neurocomputation scientists we spoke with explained how they had joined measurements with both the invasive and non-invasive neurophysiologists, worked with the involved technologies, and even participated in non-invasive EEG measurements as healthy subjects. Thus, they understand how the data is collected and, to a certain extent, empathise with patients' experiences as part of the research process. Even so, compared to neurophysiologists, different conceptualisations of patients emerge in their everyday research, which in turn are accompanied by different ethical issues or prioritisations of similar ones.

With view to data security and privacy, we made similar observations. All project parties were concerned with a suitable pre-processing and clearance of the data to maximise potential research outputs while at the same time paying attention to data privacy. Within the non-invasive neurophysiology team, researchers started discussing whether EEG data would represent a unique digital fingerprint and how to deal with this issue, especially when making the data publicly available in an Open Science sense. Albeit in a different contextual setting, the invasive neurophysiology team is also concerned with the identity of their patient, for example regarding media articles about the patient or potential adverse effects of accommodating the patient's desire to attend scientific conferences together with the team. They further discuss the possibility of recording private thoughts such as sexual arousal, fear, or anger when working with the patient, and ways of handling such incidents responsibly. Both the invasive and non-invasive neurophysiology teams emphasise the need to securely store the unprocessed data of their patients and healthy subjects. However, we noted that neurophysiologists tend to conduct these discussions mainly from the patients' perspective, underpinning them with concrete patient-related examples and reflections on possible adverse effects of data usage from their day-to-day research. Again, the patient emerges as the central focus for neurophysiologists when addressing ethical questions. Although legal regulations figure in their conversations, too, they are discussed on a par with, if not subordinate to, patient perspectives, with their sufficiency often called into question. Instead, the primary goal and benchmark is the establishment of the best possible data protection for the patient as part of their obligation to care for them.

In contrast, discussions on questions of data security among the neurocomputation groups of the network, who obtain anonymised or at least pseudonymised data after pre-processing or work with publicly available data with only minimal connection to patients or healthy subjects, are dominated by questions of suitable pre-processing, compliance with existing data protection regulations or *"approaching the problems technically"* *(PI03)*. This formulation illustrates that the neurocomputation scientists think about ethical issues on data security predominantly from the perspective of existing regulations and that they consider technical ways to comply with these regulations to be possible, workable, and desirable. Thinking about ethical questions in pragmatic and solution-oriented ways reflects the established ways in which neurocomputation scientists work and deal with scientific problems in the network. Centred in these working practices are the requirements that computer scientists face early on in their university studies and that are further cultivated throughout their academic careers. These include generating workable solutions from existing data and within given frameworks. These distinct practices ultimately shape the neurocomputation scientists' perspectives on ethical issues surrounding data security and are thus crucial to our observations and work.

Lastly, we also see a difference in how neurophysiologists and neuro-computation scientists envision the amount of data required to draw meaningful conclusions. The main concern of the non-invasive neuro-physiology team is the representativeness of their data. As described in section 3, questions of racial and ethnic bias have emerged as one of the core concerns of the team, especially against the backdrop that EEG research tends to exclude different ethnic groups, framed as a methodo-logical limitation that comes with phenotypic characteristics rather than as an equity issue (Müller et al., 2023; Webb, Etter, & Kwasa, 2022). This conversation among the neurophysiologists initiated a transnational inclu-sive dialogue about what representativeness can and should mean, and led the team to proactively promote equitable representation in neuroscientific research since then, for example at conferences. Furthermore, the group also established ties with researchers working at the interface of racism, science, and technology, including us as embedded ethicists and social scientists, to facilitate the process. The non-invasive team sees their addi-tional labour as a foundation for the long-term goal of making neu-roscience a field both more inclusive and aware of equity issues.

The question of representation also emerged within the neuro-computation teams. Similarly, to neurophysiologists, the representativeness of data is one of the essential quality criteria of the validity of their scientific work, indicating their primary socialisation as quantitative scientists. Although neurophysiologists and computer scientists at least partly share this socialisation, neurocomputation scientists approach the question of greater representativeness in a fundamentally different way. In particular, we noticed a strong conviction that enough data points would eliminate bias and that biases, for example regarding age, ethnicity or socio-economic status, would be more critical in smaller study samples than in bigger ones. Or, as one neurocomputation scientist summarised during the first internal progress meeting of the network: *"More data is always better" (R06)*. Here, fundamental confidence in the power of statistical methods arises, which is conveyed in the computer sciences as an essential part of what it means to work scientifically, cultivated through almost every aspect of their daily work practices. Consequently, technical, that is statistical, methods of dealing with issues of representativeness emerge as workable solutions that are close to the familiar routines of neurocomputation scientists and are thus adopted as their preferred way to proceed.

5. The need for situatedness of ethics and social science research in complex network structures

In this chapter, we as embedded ethicists and social scientists showed how collaborating scientists from neurophysiology and neurocomputation conceptualise ethical and social issues differently. The overarching themes that the researchers addressed were clearly similar at first glance. One of these themes, which we have explored in detail in this chapter, is data handling. This topic was raised by researchers in the neurophysiology and neurocomputation groups and includes key challenges and questions that drive their reflections on their ethical and social responsibilities. However, as we dug deeper, it became clear that the neurophysiologists were thinking from different angles and directions than the neurocomputation scientists and that certain aspects played different roles for them.

While from the outside the network appears to be a coherent consortium, a closer look reveals the complexity of the relationships and interactions between the different actors, their research practices, interests, and traditions. Our analysis points in particular to three professional socialisations interacting in the network: the medical perspective, the neurophysiological perspective, and the neurocomputational perspective drawing on mathematics and computer science. It is important to emphasise that these perspectives are not distinct but rather partially overlapping or intertwined. Even single individuals may exhibit multiple socialisations. At the same time, we have seen that specific ethical issues have emerged from these unique constellations within the network, which have been shaped by the professional socialisations of the researchers, their established work routines, positions along the data analysis process, as well as the type and frequency of patient interactions they have in their daily work. According to Merton's notion of scientific *ethos* (Merton, 1973), we observed how multiple and different *ethe* interact and sometimes conflict in our network. Specifically, three *ethe* emerged: the ethos of care among physicians, the ethos of correct knowledge among neurophysiologists, and the ethos of workable solutions among neurocomputation scientists. These interactions significantly shape the *gestalt* of our network as a constantly changing terrain of ethical and social concerns that thus affect medicine, neuroscience research, mathematics, and data science. Based on our observations, two general implications for working as embedded ethicists and social scientists within interdisciplinary contexts deserve further attention.

First, many of the ethical issues that appear overarching at first glance may emerge as increasingly multilayered, with both distinctively different and entirely congruent conceptualisations, as in our case by neurophysiologists and neurocomputation scientists. These different perspectives strongly reflect Haraway's notion of situatedness (Haraway, 1988) and Barad's (1999) apparatus metaphor. Specifically, situated ways of knowing echo in how different groups of researchers tend to frame ethical issues with view to the same topic and even within the same research network. These nuanced differences in conceptualising ethical and social issues, which nevertheless remain intertwined, do not necessarily imply a problem or a form of bias that must be mitigated. Rather, they can be seen as an invitation to accommodate the question of how contexts influence not only knowledge production processes but also the emergence and consideration of ethical and social questions. At the same time, this leads us to conclude that an overarching approach to ethical inquiry that seeks to attend to the various perspectives or interacting *ethe* cannot hold in practice given the level of complexity already emerging within our own network.

Second, the question of situatedness concerns our own research as ethicists and social scientists embedded in interdisciplinary projects. Through our research design, we adopt a position that goes beyond the individual research teams and their disciplinary imprints, that is acting in integrative ways, yet we too always see "from somewhere" (Haraway, 1988, p. 590). Consequently, embedded ethicists and social scientists need to continuously reflect on their own perspectives and to be transparent about the basis on which they select research questions, the angles from which they view material, and the limits of their own ways of knowing. This entails being aware that their research is situated in a specific context that combines constellations of actors, technologies, institutions, and infrastructures in unique configurations. To achieve this, maintaining close links with other embedded ethics initiatives can be useful, to cross-validate findings beyond specific local sites and to draw broader conclusions, both about the complex entanglements of neuroethics and AI ethics and about what embedded ethics and social science research can and should mean in practice.

6. Conclusion

Being embedded in practice as ethicists and social scientists comes with challenges, especially in an interdisciplinary context. Due to the variety of situated knowledges and different ethe, it is our task to make sense of the rising ethical and social questions by paying special attention to the respective contexts in which they have emerged. In exemplifying our work within the NEUROTECH Innovation Network, including neurocomputation scientists, neurophysiologists, and physicians, we showed how this approach currently evolves and manifests itself in practice. Although we perceived overarching similarities between AI ethics and neuroethics, the particular questions raised differ and are situated in their respective research fields.

So, how can we approach ethical and social science research if we seek to account for the nuanced differences between neuroethics and AI ethics without losing sight of their manifold intersections? Based on our findings and existing conceptualisations of situatedness, we argue for a situated approach to ethical and social scientific inquiry into the entanglements between neuroscience and AI that considers the specificities of contexts in which knowledge is produced and decisions about technological developments are made. Implementing such an approach in practice is no easy undertaking. It will require the readiness to challenge one's own thinking, a foundation of mutual understanding, and openness among all actors involved in the research. However, only by taking this additional step and considering the specific situatedness of neurophysiology and neurocomputation research without prematurely relying on their overarching similarities, will we approach meaningful ethical conclusions about controversial research and technologies in uncertain and challenging environments.

Acknowledgments

This contribution is based on the research conducted within the Innovation Network for Mental Health at TUM (NEUROTECH), which received funding via the Excellence Strategy of the Federal and State Governments. We would like to thank Prof. Dr. med. Simon N. Jacob, MD; Prof. Dr. med. Markus Ploner, MD; Prof. Dr. Daniel Rückert, PhD; Prof. Dr. Bernhard Wolfrum, PhD; Prof. Dr. med. Josef Priller, MD; Priv.-Doz. Dr. med. Jens Gempt, MD; Prof. Dr. Julijana Gjorgjieva, PhD; Prof. Dr. med. Alena Buyx, MD, and their teams for their valuable contributions and insights into their work.

References

Barad, K. (1999). *Agential realism: Feminist interventions in understanding scientific practices. The science studies reader.* London: Routledge, 1–11.

Barry, A., Born, G., & Weszkalnys, G. (2008). Logics of interdisciplinarity. *Economy and Society, 37*(1), 20–49. https://doi.org/10.1080/03085140701760841.

Berberich, N., Paredes-Acuna, N., Lipp, B., & Cheng, G. (2023). Embedding ethics into neuroengineering education: A Human-Centered Engineering Course on Neurorehabilitation. (pp. 1–4). Baltimore: IEEE. https://doi.org/10.1109/NER52421.2023.10123725.

Berger, S., & Rossi, F. (2023). AI and neurotechnology: Learning from AI ethics to address an expanded ethics landscape. *Communications of the ACM, 66*(3), 58–68. https://doi.org/10.1145/3529088.

Callard, F., Fitzgerald, D., Callard, F., & Fitzgerald, D. (2015). *States of rest: Interdisciplinary experiments. Rethinking interdisciplinarity across the social sciences and neurosciences. London*: Palgrave Macmillan, 63–78. https://doi.org/10.1057/9781137407962_5.

Callard, F., & Fitzgerald, D. (2015). *Meeting people is easy: The pragmatics of interdisciplinary collaboration. Rethinking interdisciplinarity across the social sciences and neurosciences. London*: Palgrave Macmillan, https://doi.org/10.1057/9781137407962_2.

Collins, B., & Klein, E. (2023). Invasive neurotechnology: A study of the concept of invasiveness in neuroethics. *Neuroethics, 16*(1), 11. https://doi.org/10.1007/s12152-023-09518-1.

Davis, N., & Koningsbruggen, M. (2013). "Non-invasive" brain stimulation is not non-invasive. *Frontiers in Systems Neuroscience, 7*(76), https://doi.org/10.3389/fnsys.2013.00076.

DZNE. (2022). AI, neuroscience and hardware: From neural to artificial systems and back again. https://conferences.nature.com/event/b484d933–83c0–4c87–86e0–587d9069a56d/summary Accessed 29.09.23.

Farisco, M., Evers, K. & Salles, A. On the contribution of neuroethics to the ethics and regulation of artificial intelligence. *Neuroethics 15* (4). https://doi.org/10.1007/s12152-022-09484-0.

Fiske, A., Tigard, D., Müller, R., Haddadin, S., Buyx, A., & McLennan, S. (2020). Embedded ethics could help implement the pipeline model framework for machine learning healthcare applications. *The American Journal of Bioethics, 20*(11), 32–35. https://doi.org/10.1080/15265161.2020.1820101.

Gilbert, F., Tubig, P., & Harris, A. R. (2022). Not-so-straightforward decisions to keep or explant a device: When does neural device removal become patient coercion? *AJOB Neuroscience, 13*(4), 230–232. https://doi.org/10.1080/21507740.2022.2126544.

Gilbert, F., Ienca, M., & Cook, M. (2023). How i became myself after merging with a computer: Does human-machine symbiosis raise human rights issues? *Brain Stimulation, 16*(3), 783–789. https://doi.org/10.1016/j.brs.2023.04.016.

Hall, J. G., Bainbridge, L., Buchan, A., Cribb, A., Drummond, J., Gyles, C., ... Solomon, P. (2006). A meeting of minds: Interdisciplinary research in the health sciences in Canada. *CMAJ: Canadian Medical Association Journal = Journal de l'Association Medicale Canadienne, 175*(7), 763–771.

Haraway, D. (1988). Situated knowledges: The science question in feminism and the privilege of partial perspective. *Feminist Studies, 14*(3), 575–599.

Lorach, H., Galvez, A., Spagnolo, V., et al. (2023). Walking naturally after spinal cord injury using a brain–spine interface. *Nature, 618*, 126–133. https://doi.org/10.1038/s41586-023-06094-5.

McLennan, S., Fiske, A., Celi, L. A., Müller, R., Harder, J., Ritt, K., ... Buyx, A. (2020). An embedded ethics approach for AI development. *Nature Machine Intelligence, 2*(9), 488–490. https://doi.org/10.1038/s42256-020-0214-1.

McLennan, S., Fiske, A., Tigard, D., Müller, R., Haddadin, S., & Buyx, A. (2022). Embedded ethics: A proposal for integrating ethics into the development of medical AI. *BMC Medical Ethics, 23*(6), https://doi.org/10.1186/s12910-022-00746-3.

Merton, R. K. (1973). *The sociology of science: Theoretical and empirical investigations.* University of Chicago Press ISBN: 0-226-52092-7.

Metzger, S. L., Littlejohn, K. T., Silva, A. B., et al. (2023). A high-performance neuro-prosthesis for speech decoding and avatar control. *Nature, 620*, 1037–1046. https://doi.org/10.1038/s41586-023-06443-4.

Mökander, J., & Floridi, L. (2023). Operationalising AI governance through ethics-based auditing: An industry case study. *AI and Ethics, 3*(2), 451–468. https://doi.org/10.1007/s43681-022-00171-7.

Müller, R., Ruess, A. K., Schönweitz, F. B., Buyx, A., Gil Ávila, C., & Ploner, M. (2023). Next steps for global collaboration to minimize racial and ethnic bias in neuroscience. *Nature Neuroscience, 26*(7), https://doi.org/10.1038/s41593-023-01369-6.

Parsons, T. (1951). *The social system. Glencoe.* Illinois: The Free Press.

Ploner, M., Buyx, A., Gempt, J., Gjorgjieva, J., Müller, R., Priller, J., ... Jacob, S. N. (2023). Reengineering neurotechnology: Placing patients first. *Nature Mental Health, 1*(1), https://doi.org/10.1038/s44220-022-00011-x.

Smye, S. W., & Frangi, A. F. (2021). Interdisciplinary research: Shaping the healthcare of the future. *Future Healthcare Journal, 8*(2), e218–e223. https://doi.org/10.7861/fhj.2021-0025.

Tigard, D. W., Braun, M., Breuer, S., Ritt, K., Fiske, A., McLennan, S., & Buyx, A. (2023). Toward best practices in embedded ethics: Suggestions for interdisciplinary technology development. *Robotics and Autonomous Systems, 167*, 104467. https://doi.org/10.1016/j.robot.2023.104467.

Webb, E. K., Etter, J. A., & Kwasa, J. A. (2022). Addressing racial and phenotypic bias in human neuroscience methods. *Nature Neuroscience, 25*(4), 410–414. https://doi.org/10.1038/s41593-022-01046-0.

Weh, L., & Soetebeer, M. (2023). *AI ethics and neuroethics promote relational AI discourse. Work and AI 2030.* Wiesbaden: Springer, https://doi.org/10.1007/978-3-658-40232-7_6.

Yoshida, N., Hashimoto, Y., Shikota, M., et al. (2016). Relief of neuropathic pain after spinal cord injury by brain–computer interface training. *Spinal Cord Series and Cases, 2*, 16021. https://doi.org/10.1038/scsandc.2016.21.

CHAPTER FOURTEEN

Beyond participation: Towards a community-led approach to value alignment of AI in medicine

Philipp Kellmeyer[a,b,c,*]

[a]Data and Web Science Group, School of Business Informatics and Mathematics, University of Mannheim, Mannheim, Germany
[b]Human-Technology Interaction Lab, University of Freiburg-Medical Center, Freiburg im Breisgau, Germany
[c]Institute for Biomedical Ethics and History of Medicine, University of Zurich, Zurich, Switzerland
*Corresponding author. e-mail address: philipp.kellmeyer@uni-mannheim.de

Contents

1. Introduction	242
2. Conceptualizing participation and partaking	243
2.1 Levels of participation	244
2.2 Participation as a means to increase partaking	245
2.3 Participation as a means to strengthen human capabilities	246
3. Methods of participatory research	247
3.1 Social work research	247
3.2 Educational science and pedagogy	248
3.3 Participatory approaches in design thinking and research	248
3.4 Participatory approaches in biomedical research	249
4. Current paradigms for value alignment: Ethics-by-design and embedded ethics	251
4.1 Ethics-by-design	251
4.2 Embedded ethics	251
4.3 Criticisms and the need for community inclusion	252
5. Towards community-led value alignment in medical AI development	252
5.1 Community-led research and development	253
5.2 Four key pillars of community-led development of medical AI systems	253
5.3 Context-specificity and scalability	255
6. Limits of participatory methods in medical research and AI development	256
6.1 Evidence-based medicine	256
6.2 Scalability, representativeness and bias	257
6.3 Extra burden on research participants and vulnerable communities	258
7. Conclusions	258
References	259

Developments in Neuroethics and Bioethics, Volume 7
ISSN 2589-2959, https://doi.org/10.1016/bs.dnb.2024.02.011
Copyright © 2024 Elsevier Inc. All rights are reserved, including those for text and data mining, AI training, and similar technologies.

Abstract

This paper explores the potential of community-led participatory research for value alignment in developing medical AI systems. It begins by examining conceptual aspects of participation and partaking, the current paradigms of AI development in medicine, and the emerging challenges AI technologies pose. The paper proposes a shift towards a more participatory and community-led approach to AI development, illustrated through various participatory research methods in the social sciences and design thinking. It discusses the dominant paradigms of ethics-by-design and embedded ethics in the value alignment of medical technologies, acknowledging their limitations and criticisms. A model for community-led AI development in medicine is then presented, emphasizing the importance of involving communities in all stages of the research process. The paper argues for a shift towards a more participatory and community-led approach to AI development in medicine, promising more effective and ethical medical AI systems.

1. Introduction

The accelerating innovation dynamics of artificial intelligence are on the verge of transforming clinical and healthcare research substantially (Topol, 2019). In the past thirty years, evidence-based medicine has become the paradigmatic approach to biomedical research, epitomized by randomized control trials as the gold standard for clinical translation. This research and development model is characterized by a primarily deductive, hypothesis-led approach—recent widespread concerns about replicability and the overall quality of research notwithstanding (Begley & Ioannidis, 2015).

The convergence of data science and machine learning (ML), especially with artificial neural networks for deep learning (based on deep neural networks), now enables an ever-growing range of data-driven AI approaches to research and develop AI-based medical technologies for diagnosis, predictive modeling and novel treatment strategies (Esteva et al., 2019; Kellmeyer, 2019). At the same time, however, the inherent opacity and lack of interpretability of deep neural networks and other emerging AI approaches such as generative AI, for example, large language models (LLMs), creates several, by now well characterized and discussed, challenges such as privacy (Gerber, Gerber, & Volkamer, 2018), the responsibility gap in human–AI co-actions (Kellmeyer et al., 2016), the role of trust in human–technology interaction (Schröder, Müller, Scholl, Levy-Tzedek, & Kellmeyer, 2023), bias in machine learning models and resulting discrimination (Challen et al., 2019; Londoño et al., 2022), and other issues.

Beyond participation 243

Importantly, however, most approaches to address these emerging problems of human-AI interaction still focus on finding technical solutions, such as debiasing by improving the representativeness of training datasets for ML models. What is less appreciated is the fact that AI systems, such as ChatGPT or AI-based decision-support systems in medicine, must first and foremost be understood as complex socio-technical "systems of systems" in which the hardware and software part of the system components are just one aspect, in addition to dimensions such as the materiality of the components, environmental considerations, human and interactional factors involving users and developers, normative embeddings covering ethics, human rights, laws and regulation, and other aspects. Understanding any AI system as such complex, interlocked, and embedded multi-systems suggests that simple "technofixes" will not suffice in most situations to remedy unwanted adverse effects, such as bias, discrimination, privacy violations, or exploitative and traumatizing forms of labor.[1]

Unfortunately, "AI ethics", as the currently dominant approach for framing these emerging challenges in human-AI interaction for individuals and societies that have to deal with AI effects at scale, has by now mostly become, at best, a fragmented concept and, at worst, a meaningless term for public debate, policymaking and regulation of AI.[2] As an alternative to the, more often than not, top-down approach to value alignment in AI, this chapter, therefore, focuses on participatory and partaking-oriented methods to develop AI systems in medicine, particularly community-led forms of participatory research.

2. Conceptualizing participation and partaking

Participation is a multifaceted concept that varies in depth and breadth. It can range from minimal forms, which may be used to instrumentalize participants for "ethics washing" (a phenomenon also known as *tokenism*), to more active and interactive forms of participation. This gradation of participation can be conceptualized as a "ladder of participation".

---

[1] Such as "clickwork", i.e., manual labeling of images or text performed by workers to eliminate horrific content from the training data of large-scale AI systems, such as ChatGPT (Perrigo, 2023).

[2] While there is excellent scholarship that aims to develop a rigorous and systematic approach to AI ethics, e.g., from the perspective of the philosophy of technology (Coeckelbergh, 2020), this critique is mainly addressed to empty and somewhat arbitrary calls for AI to become "more ethical" in press coverage, public debate, policy initiatives and countless self-regulatory guidelines from industry.

2.1 Levels of participation

For delineating different forms or levels of participation, the "ladder of participation" is one of the most influential conceptual models, first proposed by Sherry R. Arnstein (1969), illustrating the varying degrees of participation in a process. It provides a framework to understand the depth and breadth of involvement, ranging from minimal participation to full-fledged engagement (see Table 1).

We find no discernible participation at the first level, where individuals or communities have little to no influence over the process. This could be likened to situations in citizen policy engagement where citizens are merely *informed* about policy changes after they have been decided and implemented. They are passive recipients of information with no active role in the decision-making process. Hearings and inclusive discussions characterize the level of *preliminary participation* at which individuals are asked for their opinions, but there is no guarantee that their feedback will be incorporated into the final decision. In the context of citizen science, this might involve citizens contributing data to a research project without any influence over the research questions, data analysis, or interpretation of results.

Further up, we encounter more *active forms of participation*. Here, individuals are involved in *decision-making*, such as voting on different options. Design methods, such as *co-creation* and *co-design*, are often used for

Table 1 Levels of participation and their characteristics inspired and modified from (Wright et al., 2014).

Levels of participation	Characteristics
No participation	Instrumentalization
	Instruction
	Information
Preliminary participation	Hearing
	Inclusive discussion
Active participation	Co-design, co-creation
	Decision-making power
Beyond participation	Self-organization
	Community-led research

interactive participation. For instance, in a community planning initiative, residents might vote on different development proposals, giving them a direct say in the future of their community. Then, they might participate in design challenges in mixed teams with architects and city planners to realize the development project.

Moving *beyond participation*, in *community-led* forms of research and development, individuals lead the process and actively shape its outcomes. They work alongside professionals as equal partners in a *self-organized* manner, contributing their unique insights. In biomedical research, for instance, this would mean that patients might be involved in designing a project together with the researchers from the beginning. In AI development, this could involve users fully engaged in designing and implementing an AI system. They could contribute to defining the system's objectives, designing its features, testing its performance, and refining its functionalities based on real-world use.

This framework enables the understanding and evaluation of the level of participation in a process. It highlights the importance of moving beyond tokenistic forms of participation towards more meaningful engagement, where individuals are empowered to influence the processes and outcomes that affect their lives.

2.2 Participation as a means to increase partaking

In its simplest form, as we have seen, participation refers to participating in an activity or process. It is a fundamental aspect of democratic processes, allowing individuals or communities to have a say in matters that affect them. However, the depth and breadth of this involvement can vary significantly, from merely being informed about a decision to actively contributing to the decision-making process. On the other hand, partaking (or "Teilhabe" in German) encompasses a more profound engagement. It implies involvement in a process and a share in its outcomes, benefits, and potential drawbacks.

Through active forms of participation, partaking represents a deeper, more equitable form of engagement because individuals or communities are not merely passive recipients of decisions made by others but active contributors to the decision-making process itself. They can influence the process and its outcomes based on their unique perspectives, experiences, and insights. In this context, participation can be seen as a means to increase partaking. By actively participating in a process, individuals or communities can also influence the outcomes to align with their needs, values, and aspirations—an essential prerequisite for value alignment in innovation.

This is particularly relevant in the development of AI systems, where decisions made during the design and implementation stages can have far-reaching implications for the users of these systems. For instance, consider the development of an AI-based decision-support system for physicians. Suppose physicians, as the primary users of this system, are involved in its development. In that case, they can provide valuable insights into the usefulness of features and functionalities. They can highlight potential issues or challenges that might not be apparent to the developers and suggest solutions based on their professional expertise and experience. Participating in the development process can thus enhance their ability to provide quality patient care.

However, their involvement does not stop at the development stage. Once the system is implemented, they will continue to partake in its outcomes by using it in their daily practice and giving feedback on its performance. They experience the system's benefits first-hand, such as increased efficiency or improved diagnostic accuracy. They might also face any drawbacks or challenges that come with its use, such as technical issues or changes in their workflow. Their ongoing feedback can then refine and improve the system further. Active participation, therefore, allows individuals and communities to influence processes and share in their outcomes to increase partaking. This potentially leads to better, more inclusive decisions and fosters a sense of ownership and responsibility among those involved.

2.3 Participation as a means to strengthen human capabilities

As we continue to navigate the complex landscape of medical AI development, these concepts will be crucial in ensuring that the systems we develop are technically sound and socially and ethically responsible. The capability approach, developed by Martha Nussbaum (2002) and Amartya Sen (2005), argues that the key to a good life lies in the ability to function in various aspects of life, such as health, education, and political participation. From this perspective, participation is not just a means to an end but an end in itself. It provides an opportunity for increased participation in life's cultural, economic, political, and social dimensions.

Increasing participation can thus help to strengthen capabilities. For instance, involving patients in designing a healthcare AI system can empower them to manage their health better, enhancing their capability to function in the health dimension of life. Understanding participation as a gradual process and leveraging it as a tool to increase partaking and strengthen capabilities can thus provide valuable insights for developing medical AI systems.

3. Methods of participatory research

While the scope of this chapter does not allow to give a complete panoramic overview and academic history of participatory research methods and models, some examples from the social sciences and design research can be particularly instructive to understand why participation might be a promising approach in medical AI development.

There is now a wide range of methodologies and approaches in participatory research in the social sciences that share many similarities yet are often very context-specific, thereby eluding a more generalized taxonomy of methods. As a methodological snapshot, we find approaches such as collaborative inquiry (Bray, 2000), cooperative inquiry (Heron & Reason, 2006), and action research (Bradbury, 2015) that are themselves now becoming diversified and applied in different fields and domains.

Broadly construed, participatory research is an approach that emphasizes collaboration and partnership between the researcher and the community being studied. Various participation models have enjoyed longstanding attention, particularly in the social sciences, which often work closely with specific groups such as social workers or educational scientists. All these approaches in the social sciences are characterized by collaboration, inclusivity, and a commitment to social change and challenge traditional research dynamics by offering new ways of knowing and learning.

3.1 Social work research

In social work research, participatory research is crucial in understanding and addressing complex social issues (Branom, 2012; Flanagan, 2020). This approach is particularly effective because it involves those directly experiencing the issues, thereby allowing for the inclusion of often marginalized voices and promoting social change.

Common methods used in participatory social work research include focus groups, interviews, and community meetings. These methods foster open dialog and shared decision-making between researchers and participants, ensuring the research process is collaborative and inclusive.

For instance, consider a social work initiative to improve mental health support for homeless communities. In this context, participatory research might involve organizing focus groups with participants from the homeless community to understand their particular mental health needs. These focus groups could provide a safe space for individuals to share their experiences, which might otherwise remain unheard. In addition to focus groups, one-on-one interviews could be conducted to delve deeper into individual experiences and perspectives. These interviews could explore the barriers to accessing mental health services, the types of support individuals find most helpful, and suggestions for improving existing services. Community meetings could also be organized to facilitate broader discussions and collective decision-making. These meetings could involve homeless community members, mental health professionals, policymakers, and other stakeholders. Together, they could discuss the findings from the focus groups and interviews, brainstorm potential solutions, and make decisions about the direction of the mental health support initiative.

Through these participatory research methods, social workers can ensure that their efforts to improve mental health support for homeless communities are grounded in the lived experiences of those they aim to help. This approach enhances the relevance and effectiveness of their work and empowers the individuals and communities involved, fostering a sense of ownership and agency.

3.2 Educational science and pedagogy

In the field of educational science, participatory research methods are used to involve students, teachers, parents, and other stakeholders in the research process. This approach recognizes that these individuals are experts in their experiences and can provide valuable insights into educational practices and policies. Methods used in participatory educational research often include collaborative curriculum development, student-led conferences, and participatory action research. Participatory research in pedagogy involves learners in a collaborative process of knowledge creation (Andersen & Ponti, 2014). Approaches such as collaborative learning, project-based learning, and learner-led assessments, grounded in theories of participatory learning in childhood development such as participatory learning theory (Hedges & Cullen, 2012), challenge traditional power dynamics in education.

3.3 Participatory approaches in design thinking and research

Design thinking and research have evolved to incorporate participatory approaches, fostering a more inclusive and collaborative process. Some important methods and approaches in design research, especially in human-technology interaction research, are co-design, co-creation, rapid prototyping, techno-mimesis, and design fictions (Ammon, 2020).

Co-design taps into the collective creativity of designers and users throughout the design cycle. It is a democratic approach that values the input of all stakeholders, recognizing that everyone has something to contribute to the design process. *Co-creation*, on the other hand, is a broader concept that involves collaboration between different parties, e.g., designers, users, and stakeholders, to produce a mutually valued outcome. It extends beyond the design phase to include the implementation and use of the designed artifact.

Several methods facilitate co-design and co-creation. *Techno-mimesis* is one such method (Dörrenbächer, Löffler, & Hassenzahl, 2020). It involves mimicking or simulating technology with everyday materials in interactive roleplays to understand its potential implications and uses. In medical robotics, for instance, techno-mimesis could involve roleplaying exercises where humans are transformed into a design-fiction version of the robot with everyday materials, mimicking the actions of a robotic surgeon to explore new possibilities for surgical procedures.

Rapid prototyping is another method used in co-design. It involves quickly creating a functional model of a product to test its feasibility and effectiveness. This method allows for iterative testing and refinement of the design based on user feedback. For example, in developing a virtual reality system for pain management, rapid prototypes could be used to test different virtual environments and gauge their effectiveness in distracting patients from pain.

Lastly, *design fictions* are speculative designs that provoke discussion and debate about the future. They are not intended to be practical solutions but to stimulate imagination and critical thinking. In the realm of AI in medicine, a design fiction could be a narrative or scenario that explores the potential consequences of AI diagnostic systems, both positive and negative.

These participatory design methods offer promising avenues for developing medical technologies. Involving many stakeholders in the design process can ensure that the resulting technologies are technically sound and assessed regarding their inherent or complex ethical and social risks.

3.4 Participatory approaches in biomedical research

Participation in biomedical research recognizes the unique insights individuals with lived experience, such as patients, can bring. This is because lived experience offers a less filtered, first-hand perspective of the patient's journey, providing unique insights that cannot be obtained from indirect or mediated sources. These insights are crucial in shaping patient-centered care, enhancing doctor-patient communication, and informing medical education. Therefore, the integration of lived experience into biomedical research is not just beneficial but essential for the advancement of healthcare.

Example 1: Participatory research in diabetes management.

One example of participatory research in medicine is the development of self-management interventions for patients with diabetes (Campbell, Yan, & Egede, 2020). Diabetes is a chronic condition that requires regular blood glucose monitoring, medication, diet, and exercise. Patients are often the primary managers of their condition, making them experts in their care. In many participatory research settings, a team of researchers collaborates with a group of patients with diabetes to develop a new self-management intervention. The patients are involved in all stages of the research process, from defining the research question to interpreting the results. For instance, patients might share their diabetes management experiences, including the challenges and strategies they find helpful. These insights then inform the development of the intervention, ensuring it is relevant and responsive to patients' needs. The patients might also be involved in testing the intervention providing feedback on its usability and effectiveness. Their input could lead to further refinements, resulting in a final product that is evidence-based and grounded in the lived experiences of those it is designed to help.

Example 2: Participatory research in mental health services.

Another example of participatory research in medicine is evaluating mental health services (Schneider, 2010; Stacciarini, Shattell, Coady, & Wiens, 2011). Mental health services can vary widely in their approach and effectiveness, and understanding the experiences of those who use these services is crucial for their improvement. In participatory mental health research, researchers typically collaborate with a group of mental health service users regarding a particular intervention. The service users are involved in designing the evaluation, deciding what aspects of the intervention to focus on, and developing the data collection tools. For instance, service users might choose to focus on the accessibility of the service, the quality of the therapeutic relationship, or the outcomes of the treatment. They might develop a survey or interview guide to collect data on these aspects, drawing on their own experiences to ensure the data collection tools are comprehensive and relevant. The service users may also analyze the data and interpret the results. Their insights could highlight areas of strength in the service and areas that need improvement.

These examples illustrate how participatory research in medicine offers a powerful approach to ensuring that research is relevant, responsive, and respectful to those it is designed to help. By actively involving patients and healthcare providers in research, we can generate insights grounded in lived experience and professional expertise leading to more effective and equitable healthcare solutions.

4. Current paradigms for value alignment: Ethics-by-design and embedded ethics

The rapid advancement of technology, particularly in medicine, has necessitated the integration of ethical considerations into the design and development process. Two dominant paradigms have emerged in response to this need: ethics-by-design and embedded ethics.

4.1 Ethics-by-design

The ethics-by-design paradigm originated from the broader concept of value-sensitive design, an approach that advocates for considering human values throughout the technological design process (Tonkinwise, 2004; Van Der Hoven & Manders-Huits, 2020). This approach has been applied across various fields, from information systems to architecture, and has now found its way into medical technology (Ienca, Wangmo, Jotterand, Kressig, & Elger, 2017; Kellmeyer, Biller-Andorno, & Meynen, 2019).

Ethics-by-design goes beyond mere compliance with ethical guidelines or regulations. It involves proactively identifying and addressing potential ethical issues during the design process. This approach ensures that the resulting technology functions effectively and aligns with societal values and norms. In medical technology, ethics-by-design could involve considerations around patient privacy when designing electronic health records or fairness and transparency when developing AI diagnostic tools (Price & Cohen, 2019). For instance, a team developing a new AI system for diagnosing skin cancer might incorporate ethics-by-design by ensuring the algorithm is trained on a diverse dataset, thus minimizing potential bias and improving the system's fairness.

4.2 Embedded ethics

Embedded ethics, on the other hand, refers to integrating ethical considerations into professional practices and processes, for example, in clinical ethics or the development and use of technology (McLennan et al., 2022). This approach recognizes that ethical issues are not just design problems to be solved but are deeply intertwined with the contexts in which technology is used.

In clinical ethics counseling, for example, embedded ethics might involve regular discussions about ethical issues in team meetings or the inclusion of ethicists in medical wards with frequent needs for counseling, such as intensive care units. This approach ensures that ethical considerations are not an afterthought but are central to the team's everyday work.

When developing medical technologies, embedded ethics could involve regular ethical reviews throughout the development process or the inclusion of diverse perspectives, in decision-making processes, for example from patients, healthcare providers, and ethicists. For example, a team developing a social robot for elder care might incorporate embedded ethics by involving potential end-users (e.g., elderly individuals, caregivers) in the design process, ensuring the robot's functionality and behavior align with the needs and values of those it is designed to assist. The chapters in this volume by McLennan, Willem and Fiske and Schönweitz, Ruess and Müller respectively also provide a detailed discussion of this approach.

4.3 Criticisms and the need for community inclusion

Despite the potential benefits of ethics-by-design and embedded ethics, these approaches are not without their criticisms (Dignum et al., 2018). First, in many instances, they amount to top-down approaches that impose a set of predefined ethical principles onto the design process, potentially overlooking the diverse values and needs of different user groups. Second, they often rely on abstract moral principles that may not translate well into concrete design decisions. Finally, these approaches can sometimes lead to a check-box mentality, where ethical considerations are seen as items to be ticked off a list rather than integral aspects of the design process (Kellmeyer, 2018; Keymolen & Taylor, 2023).

These criticisms highlight the need for a more community-led approach to AI development in medicine. By involving communities in the design process, we can ensure that the resulting technologies are technically sound and genuinely responsive to the needs and values of those they are designed to assist. As we move towards a more community-led approach to AI in medicine, these participatory research methods offer valuable insights and models for practice. The following section will explore how a community-led approach can be applied to the value alignment of AI systems in medicine.

5. Towards community-led value alignment in medical AI development

The development of medical AI systems is a complex process that requires technical expertise and a deep understanding of the needs and values of the communities these systems are designed to serve. However, while participatory paradigms are now increasingly used to align medical technologies with user needs based on ergonomic features or other useability metrics, community-led forms of engagement are not yet widely used. In this section, I will argue that going beyond participation by implementing models of community-led research can contribute to the value alignment in developing medical AI systems.

5.1 Community-led research and development

As outlined above, community-led research is characterized by the active involvement of community members in all stages of the research process, from defining the research question to interpreting the results. This approach recognizes the unique insights that community members can bring to the research process and seeks to empower them as co-researchers.

One way to conceptualize the degree of community participation and engagement in research is through the levels of participation illustrated in Table 1. At the most basic level are traditional research approaches where community members are merely subjects of research. Community members become increasingly involved in the research process in more participatory constellations, eventually becoming equal partners. Moving beyond participation, communities lead the research, working with experts and other stakeholders.

For instance, consider the development of an AI-based implantable system for blood sugar management in diabetes. In a traditional research approach, patients might be asked to trial the system and provide feedback. However, in a community-led approach, patients would (co-)design the research questions, interpret the data, and disseminate the results. They might, for example, contribute ideas on how to make the system more user-friendly or suggest new features that would be particularly useful for managing their condition.

But what key pillars and conditions must be fostered to enable such community-led efforts for value alignment of medical AI systems?

5.2 Four key pillars of community-led development of medical AI systems

Implementing community-led research, particularly in co-creating medical AI systems, requires careful consideration of several key pillars and preconditions. These foundational elements ensure the research process is highly participatory in practice, leading to outcomes that genuinely reflect the community's needs and values.

Pillar 1: Community engagement and trust building.

The first pillar of successful community-led research is robust community engagement and trust building. This involves establishing strong relationships with community members based on mutual respect and understanding. In the context of developing a medical AI system, this might involve regular meetings with potential users of the system, such as patients, healthcare providers, and caregivers. These meetings provide an opportunity to discuss the project, address any concerns, and build a shared vision for the AI system.

Pillar 2: Inclusive and equitable participation.

The second pillar is ensuring inclusive and equitable participation. This means that all community members have an opportunity to contribute to the research process, regardless of their background or expertise. For instance, when developing an AI-based diagnostic tool, it is important to include medical professionals and patients who can provide valuable insights based on their lived experiences. This inclusivity ensures that the resulting AI system is responsive to the needs of all potential users.

Pillar 3: Capacity building.

The third pillar is capacity building. This involves providing community members with the skills and knowledge to actively participate in the research process. In AI development, this might involve workshops or training sessions on basic AI concepts, data analysis techniques, or ethical considerations in AI. By building capacity, community members are empowered to contribute meaningfully to the research process, enhancing the quality and relevance of the research outcomes.

Pillar 4: Iterative and reflective learning.

The fourth pillar is adopting an iterative and reflective process. Community-led research is not a linear process but a cyclical one, where insights gained at one stage inform the next. For example, feedback from users on a prototype of a medical AI system might lead to revisions in the system's design. Regular reflection on the research process and outcomes is crucial for continuous learning and improvement.

Therefore, implementing community-led research in developing medical AI systems requires a strong commitment to community engagement, inclusive participation, capacity building, and iterative learning. However, two main challenges for such highly engaged forms of co-development are context-specificity and scalability (see also **Section 6**).

5.3 Context-specificity and scalability

While community-led research is inherently context-specific, it can also be designed to be scalable to ensure representativeness. This involves balancing the need for deep, context-specific insights with generalizable findings in other contexts. This balance is crucial in AI development, where the goal is to create adaptable and effective systems across a wide range of scenarios and populations.

One way to achieve this balance is through a multi-level approach to community-led research. At the local level, in-depth, context-specific research is conducted with a particular community. This might involve focus groups, interviews, or participatory workshops where community members can share their experiences, needs, and ideas. The aim is to gain a deeper understanding of the specific context and how it might influence the use and impact of the AI system.

The findings from this local research are then fed into a higher level of analysis, combined with findings from other local studies. This could involve a meta-analysis or a cross-case analysis, where common themes and patterns are identified across different communities. This process allows for the extraction of broader insights not tied to a specific context, thereby increasing the scalability of the findings.

Moreover, this approach acknowledges differences between communities, which can be important for tailoring medical AI systems to specific contexts. For instance, an AI system designed for urban healthcare providers might need to be adapted for use in rural communities, where healthcare resources and infrastructure might differ. In implementing this multi-level approach, ensuring the process is transparent and inclusive is important. Communities should be involved not only in the local research but also in the higher-level analysis. This could involve sharing preliminary findings with the communities and seeking their input on interpreting the results. This ensures that the research process is participatory and empowering, giving communities a sense of ownership over the research and its outcomes.

Example: Developing an AI-based decision-support system.

A mixed-methods research design that involves participatory methods from qualitative research can be particularly useful for developing medical AI systems. This approach combines quantitative methods, which help test hypotheses and establish generalizable findings, with qualitative methods, useful for exploring complex phenomena in depth.

Consider, for example, a study on developing an AI-based decision-support system for automated EEG analysis in neurology. The study might start with a quantitative phase, where existing EEG data is used to train and test the AI system. This would provide a measure of the system's accuracy and efficiency.

The study would then move into a qualitative phase, where neurologists, patients, and other stakeholders are involved in a participatory process to refine the system. This might include focus groups or workshops where participants can provide feedback on the system, suggest improvements, and discuss ethical considerations. The insights from this qualitative phase would then be used to refine the system further, resulting in a medical AI system that is technically sound and aligned with the needs and values of the community it is designed to serve.

6. Limits of participatory methods in medical research and AI development

While participatory methods have great potential to democratize medical AI development, they are not without their limitations and criticisms. This section will explore some of these issues, focusing on evidence-based medicine, scalability, replicability, and the extra burden participatory research can place on individuals.

6.1 Evidence-based medicine

One of the main criticisms of participatory methods in medical research is that these methods might not meet the rigorous empirical requirements of evidence-based medicine. Evidence-based medicine prioritizes using high-quality, quantitative evidence from randomized controlled trials and systematic reviews in making healthcare decisions, and it is difficult to include multidimensional qualitative data in this framework as qualitative methods and experiential knowledge can sometimes be seen as less rigorous or reliable.

For example, the emerging field of meta-science research has spotlighted deep-seated methodological problems of the scientific process in many fields, biomedical research included, which is often summarized under the notion of a "replicability crisis". Replicability, or the ability to reproduce research findings, is a cornerstone of scientific research. This crisis refers to the growing concern that many scientific studies, particularly in fields like psychology and biomedical research, are difficult to replicate, casting doubt on the reliability of their findings. Factors contributing to this crisis include publication bias, p-hacking, lack of transparency in methodology, and insufficient oversight in peer review. The reproducibility crisis has prompted calls for reforms in scientific practices to enhance the reliability and integrity of scientific research (Anderson, Eijkholt, & Illes, 2013; Kellmeyer, 2017). However, participatory methods, emphasizing context-specific research and collaboration with specific communities, can be difficult to replicate. This can make building a cumulative body of knowledge challenging, as findings from one study may not directly apply to other contexts or populations. However, it is important to note that different types of evidence are valuable for answering other questions, and participatory methods can provide important insights that quantitative methods might miss.

6.2 Scalability, representativeness and bias

Scalability indeed poses a significant challenge for participatory methods. While excellent for in-depth, context-specific research, these methods can be challenging to scale up to larger populations. This is particularly relevant in AI development, where algorithms often must be trained on large datasets to perform effectively and to be robust and representative.

For instance, consider the development of an AI-based app for mental health support. The app might use machine learning algorithms to provide personalized therapeutic interventions—so-called ecological momentary interventions—based on user input. In the development phase, a participatory approach might involve working closely with a small group of users to understand their needs and preferences and tailor the app's features accordingly. However, scaling this process up to a larger population can be challenging. The insights gained from a small, potentially non-representative group of users with lived experience of mental health issues might not generalize to the broader population of app users. This could lead to a system that works well for the initial group of users but performs poorly for others. Moreover, working with smaller, potentially non-representative communities may also heighten the risk of introducing bias at the level of training data. Suppose the algorithm is trained primarily on data from a few users. In that case, it might perform less effectively for users from different demographic groups or with different mental health conditions.

While multi-level approaches can help address this issue, as discussed above, scalability remains a significant challenge for participatory AI development. It requires careful planning and methodological rigor to ensure that the insights gained from participatory methods can effectively translate into AI systems that work well for diverse users.

6.3 Extra burden on research participants and vulnerable communities

Finally, participatory methods can burden individuals, particularly those already vulnerable, for example, through pre-existing health conditions (Kellmeyer, 2020). Participatory research often involves a significant time commitment from participants, who are asked to contribute to various stages of the research process. This can be particularly challenging for individuals dealing with health issues, financial instability, or other forms of vulnerability (Herzog, Kellmeyer, & Wild, 2021). Furthermore, the emotional labor involved in sharing personal experiences and insights can also be significant. It is crucial for researchers using participatory methods to be aware of these potential burdens and to take steps to mitigate them, such as providing appropriate compensation for participants' time and emotional labor.

7. Conclusions

This chapter has explored the potential of community-led participatory research for value alignment in developing medical AI systems. It began by examining the current paradigms of AI development in medicine, which are characterized by a primarily deductive and hypothesis-led approach. The paper then highlighted the emerging challenges posed by AI technologies' inherent opacity and lack of interpretability, such as privacy concerns, the responsibility gap in human-AI co-actions, and bias in machine learning models.

In response to these challenges, the paper proposed a shift towards a more participatory and community-led approach to AI development. This approach was illustrated through various participatory research methods in the social sciences, particularly social work, educational science, and pedagogy. The paper also discussed participatory approaches and models in design thinking and research, including the concepts of co-design and co-creation and the methods of techno-mimesis, rapid prototyping, and design fictions.

The dominant paradigms of ethics-by-design and embedded ethics were then examined, focusing on their applications in the development of medical technologies. The community-led approach for value alignment presented here, emphasizes the importance of involving communities in all stages of the research process, from defining the research question to interpreting the results. It also highlights the need to balance context-specific research and scalability to ensure representativeness.

The potential of all these paradigms for value alignment for ensuring that medical AI systems are technically sound and socially and ethically responsible was highlighted. However, we have to acknowledge the limitations and criticisms of these approaches, including issues related to evidence-based medicine, scalability, replicability, and the extra burden they can place on individuals.

References

Ammon, S. (2020). *Design methods and validation. The routledge handbook of the philosophy of engineering*. Routledge, 315–327. https://doi.org/10.4324/9781315276502-28.

Andersen, R., & Ponti, M. (2014). Participatory pedagogy in an open educational course: Challenges and opportunities. *Distance Education, 35*(2), 234–249. https://doi.org/10.1080/01587919.2014.917703.

Anderson, J. A., Eijkholt, M., & Illes, J. (2013). Ethical reproducibility: Towards transparent reporting in biomedical research. *Nature Methods, 10*(9), 843.

Arnstein, S. R. (1969). A ladder of citizen participation. *Journal of the American Institute of Planners, 35*(4), 216–224. https://doi.org/10.1080/01944366908977225.

Begley, C. G., & Ioannidis, J. P. A. (2015). Reproducibility in science. *Circulation Research, 116*(1), 116–126. https://doi.org/10.1161/CIRCRESAHA.114.303819.

Bradbury, H. (2015). *The Sage handbook of action research*. SAGE.

Branom, C. (2012). Community-based participatory research as a social work research and intervention approach. *Journal of Community Practice, 20*(3), 260–273. https://doi.org/10.1080/10705422.2012.699871.

Bray, J. N. (2000). *Collaborative inquiry in practice: Action reflection, and making meaning*. SAGE.

Campbell, J. A., Yan, A., & Egede, L. E. (2020). Community-based participatory research interventions to improve diabetes outcomes: A systematic review. *The Diabetes Educator, 46*(6), 527–539. https://doi.org/10.1177/0145721720962969.

Challen, R., Denny, J., Pitt, M., Gompels, L., Edwards, T., & Tsaneva-Atanasova, K. (2019). Artificial intelligence, bias and clinical safety. *BMJ Quality & Safety, 28*(3), 231–237. https://doi.org/10.1136/bmjqs-2018-008370.

Coeckelbergh, M. (2020). *AI ethics*. The MIT Press, ⟨https://www.gbv.de/dms/bowker/toc/9780262538190.pdf⟩.

Dignum, V., Baldoni, M., Baroglio, C., Caon, M., Chatila, R., Dennis, L., ... de Wildt, T. (2018). Ethics by design: Necessity or curse? *Proceedings of the 2018 AAAI/ACM conference on AI, ethics, and society*, 60–66. https://doi.org/10.1145/3278721.3278745.

Dörrenbächer, J., Löffler, D., & Hassenzahl, M. (2020). Becoming a robot—Overcoming anthropomorphism with techno-mimesis. *Proceedings of the 2020 CHI conference on human factors in computing systems*, 1–12. https://doi.org/10.1145/3313831.3376507.

Esteva, A., Robicquet, A., Ramsundar, B., Kuleshov, V., DePristo, M., Chou, K., ... Dean, J. (2019). A guide to deep learning in healthcare. *Nature Medicine, 25*(1), 24–29. https://doi.org/10.1038/s41591-018-0316-z.

Flanagan, N. (2020). Considering a participatory approach to social work—Service user research. *Qualitative Social Work, 19*(5–6), 1078–1094. https://doi.org/10.1177/1473325019894636.

Gerber, N., Gerber, P., & Volkamer, M. (2018). Explaining the privacy paradox: A systematic review of literature investigating privacy attitude and behavior. *Computers & Security, 77*, 226–261. https://doi.org/10.1016/j.cose.2018.04.002.

Hedges, H., & Cullen, J. (2012). Participatory learning theories: A framework for early childhood pedagogy. *Early Child Development and Care, 182*(7), 921–940. https://doi.org/10.1080/03004430.2011.597504.

Heron, J., & Reason, P. (2006). The practice of co-operative inquiry: Research 'with' rather than 'on' people. *Handbook of action research: Concise paperback edition*, 144–154.

Herzog, L., Kellmeyer, P., & Wild, V. (2021). Digital behavioral technology, vulnerability and justice: Towards an integrated approach. *Review of Social Economy, 0*(0), 1–22. https://doi.org/10.1080/00346764.2021.1943755.

Ienca, M., Wangmo, T., Jotterand, F., Kressig, R. W., & Elger, B. (2017). Ethical design of intelligent assistive technologies for dementia: A descriptive review. *Science and Engineering Ethics*, 1–21. https://doi.org/10.1007/s11948-017-9976-1.

Kellmeyer, P. (2017). Ethical and legal implications of the methodological crisis in neuroimaging. *Cambridge Quarterly of Healthcare Ethics: CQ: The International Journal of Healthcare Ethics Committees, 26*(4), 530–554. https://doi.org/10.1017/S096318011700007X.

Kellmeyer, P. (2018). Big brain data: On the responsible use of brain data from clinical and consumer-directed neurotechnological devices. *Neuroethics*. https://doi.org/10.1007/s12152-018-9371-x.

Kellmeyer, P. (2019). Artificial intelligence in basic and clinical neuroscience: Opportunities and ethical challenges. *Neuroforum, 25*(4), 241–250. https://doi.org/10.1515/nf-2019-0018.

Kellmeyer, P. (2020). Digital vulnerability: A new challenge in the age of super-convergent technologies. *Bioethica Forum, 12*(1/2), 60–62.

Kellmeyer, P., Biller-Andorno, N., & Meynen, G. (2019). Ethical tensions of virtual reality treatment in vulnerable patients. *Nature Medicine, 1*. https://doi.org/10.1038/s41591-019-0543-y.

Kellmeyer, P., Cochrane, T., Müller, O., Mitchell, C., Ball, T., Fins, J. J., & Biller-Andorno, N. (2016). The effects of closed-loop medical devices on the autonomy and accountability of persons and systems. *Cambridge Quarterly of Healthcare Ethics: CQ: The International Journal of Healthcare Ethics Committees, 25*(4), 623–633. https://doi.org/10.1017/S0963180116000359.

Keymolen, E., & Taylor, L. (2023). Data ethics and data science: An uneasy marriage? In W. Liebregts, W.-J. Van Den Heuvel, & A. Van Den Born (Eds.). *Data science for entrepreneurship: Principles and methods for data engineering, analytics, entrepreneurship, and the society* (pp. 481–499). Springer International Publishing. https://doi.org/10.1007/978-3-031-19554-9_20.

Londoño, L., Hurtado, J. V., Hertz, N., Kellmeyer, P., Voeneky, S., & Valada, A. (2022). Fairness and bias in robot learning. *arXiv*. https://doi.org/10.48550/arXiv.2207.03444.

McLennan, S., Fiske, A., Tigard, D., Müller, R., Haddadin, S., & Buyx, A. (2022). Embedded ethics: A proposal for integrating ethics into the development of medical AI. *BMC Medical Ethics, 23*(1), 6. https://doi.org/10.1186/s12910-022-00746-3.

Nussbaum, M. (2002). Capabilities and social justice. *International Studies Review, 4*(2), 123–135. https://doi.org/10.1111/1521-9488.00258.

Perrigo, B. (2023). Exclusive: The $2 per hour workers who made ChatGPT safer. *Time*. ⟨https://time.com/6247678/openai-chatgpt-kenya-workers/⟩.

Price, W. N., & Cohen, I. G. (2019). Privacy in the age of medical big data. *Nature Medicine, 25*(1) Article 1. https://doi.org/10.1038/s41591-018-0272-7.

Schneider, B. (2010). *Hearing (our) voices: Participatory research in mental health*. University of Toronto Press.

Schröder, I., Müller, O., Scholl, H., Levy-Tzedek, S., & Kellmeyer, P. (2023). Can robots be trustworthy? *Ethik in Der Medizin, 35*(2), 221–246. https://doi.org/10.1007/s00481-023-00760-y.

Sen, A. (2005). Human rights and capabilities. *Journal of Human Development, 6*(2), 151–166. https://doi.org/10.1080/14649880500120491.

Stacciarini, J.-M. R., Shattell, M. M., Coady, M., & Wiens, B. (2011). Review: Community-based participatory research approach to address mental health in minority populations. *Community Mental Health Journal, 47*(5), 489–497. https://doi.org/10.1007/s10597-010-9319-z.

Tonkinwise, C. (2004). Ethics by design, or the ethos of things. *Design Philosophy Papers,* *2*(2), 129–144. https://doi.org/10.2752/144871304×13966215067994.

Topol, E. J. (2019). High-performance medicine: The convergence of human and artificial intelligence. *Nature Medicine, 25*(1), 44–56. https://doi.org/10.1038/s41591-018-0300-7.

Van Der Hoven, J., & Manders-Huits, N. (2020). *Value-sensitive design. The ethics of information technologies.* Routledge, 329–332. ⟨https://www.taylorfrancis.com/chapters/edit/10.4324/9781003075011-23/value-sensitive-design-jeroen-van-der-hoven-noemi-manders-huits⟩.

Wright, M., Kilian, H., Block, M., Von Unger, H., Brandes, S., Ziesemer, M., ... Rosenbrock, R. (2014). Partizipative Qualitätsentwicklung: Zielgruppen in alle Phasen der Projektgestaltung einbeziehen. *Das Gesundheitswesen, 77*(S 01), S141–S142. https://doi.org/10.1055/s-0033-1347268.

Epilogue: Harmonizing the ethical symbiosis of brains and machines

Marcello Ienca[a,b]
[a]Institute for History and Ethics of Medicine, School of Medicine and Health, Technical University of Munich (TUM), Germany
[b]College of Humanities, EPFL, Switzerland

As neuroscience and artificial intelligence (AI) advance, the vision of a future where intelligent machines record, complement or enhance human cognitive capacities is rapidly becoming a tangible reality (Ienca and Ignatiadis, 2020). This fusion, while holding enormous promise, also beckons the scientific community to steward a new ethical paradigm—one that harmonizes the symbiosis of brains and machines.

The contributions in this book attest to how this growing symbiosis of biological and artificial cognitive systems forces us to navigate the intricate ethical issues no longer in silos. Historically, the ethical issues arising at the intersection of neuroscience and AI have been approached in a compartmentalized manner—neuroethics, bioethics, and AI ethics have often been addressed as distinct domains of applied ethics. Each field has developed its own frameworks and guidelines, focused on the specific challenges and questions pertinent to its scope. Neuroethics has grappled with the implications of understanding and interfacing with the brain. Bioethics with the broader implications for life and organic systems. AI ethics has been dealing with the burgeoning intelligence and alleged agency of synthetic entities.

However, as the boundaries between these domains become increasingly permeable and the interactions more complex, such siloed approaches appear increasingly inadequate. The convergence of AI and neuroscience is giving rise to novel ethical quandaries that span across these traditional divides, revealing the need for a harmonized and holistic framework. This framework must integrate the insights and principles from each domain to address the multifaceted ethical landscape that is emerging. It should account for the shared concerns of cognitive liberty, mental integrity, privacy, the dual-use of technology, and the enhancement debate, among others.

In computational neurology and psychiatry, AI's prowess in integrating neural data to personalize brain and mental health care epitomizes the double-edged sword of technological advancement. While the potential for tailored treatment and diagnosis is unprecedented, it is accompanied by a constellation of ethical concerns. Mental privacy, i.e., the opacity of the individual's mental landscape, is intimately connected to the transparency of algorithms used to decode mental information. The advent of digital phenotyping, leveraging ubiquitous technology like smartphones, brings the ethics of consent and the potential for neuromarketing into the fore, challenging us to safeguard the individual against the commodification of their cognitive and emotional patterns (Martinez-Martin, Insel, Dagum, Greely, & Cho, 2018).

The ethical considerations of anthropomorphism in AI, particularly in social chatbots based on large language models, underscore the complexity of human–AI relationships (Salles, Evers, & Farisco, 2020). These digital inter-locutors, devoid of corporeal form (unlike robots) but designed to simulate humanlike interaction, blur prima facie the lines between technology and life. This new form of human–machine interaction makes traditional Turing tests obsolete and turbocharges older ethical questions about deception. The ontological and pragmatic approaches to AI's anthropomorphization demand a more sophisticated ethical discourse—one that navigates the psychological impact of AI systems and scrutinizes the conceptual assumptions of agency and intentionality we attribute to such systems.

The sense of agency (SoA), a cornerstone of moral responsibility, finds itself under the magnifying glass in the AI era. Our interactions with intel-ligent systems are recalibrating our perceptions of control and responsibility. As AI systems are imbued with greater autonomy, they challenge the tra-ditional boundaries of SoA, prompting us to reconsider our ethical frame-works in light of shared agency with non-human entities.

The prospect of artificial consciousness also raises profound ethical dilemmas. It forces us to question the essence of consciousness and the conditions under which an artificial system might be considered a moral subject. This inquiry not only demands a theoretical and technical exploration but also an ethical one that is inclusive and non-anthropocentric, recognizing the moral worth of all conscious beings—biological or synthetic.

The integration of AI into brain-computer interfaces (BCIs) exemplifies the delicate balance between enhancement and clinical intervention. As BCIs evolve into more complex, bidirectional systems capable of aug-menting human experience and treating neurological disorders, they also raise ethical challenges related to mental integrity, hybrid identity, and the potential for misuse of cognitive processing. The ethical stewardship of such technologies is paramount to ensure they enhance human capabilities without compromising the individual's sovereignty over their mental and neurological self. The concept of algorithmic regulation offers a novel perspective on the governance of BCIs and other neurotechnologies. As direct-to-consumer neurotechnologies permeate global markets, their capability to shape behavior and influence societal norms through data-driven mechanisms requires a robust regulatory and ethical framework that addresses questions of autonomy, identity, and the right to cognitive liberty (Ienca et al., 2022).

The interplay of hot and cold cognition in both humans and AI also presents a new frontier for ethical exploration. As AI systems begin to emulate the nuanced social and emotional intelligence of humans, the ethical implications of such capabilities come into focus. The integration of hot cognitive processes in AI necessitates a conscientious approach that considers the ethical ramifications of more humanlike AI–human interactions.

Finally, the philosophical debates surrounding the extended mind and cognitive artifacts enrich our understanding of the cognitive entwinement with technology. These discussions illuminate the ethical nuances of such an integration. Furthermore, they call for a reassessment of the ethical status of cognitive tools that become extensions of our mental processes.

To produce this unified ethical paradigm that spans across neuroscience and AI, novel approaches to ethical inquiry will be needed. The embedded ethics approach is emerging as a privileged modus operandi to assess AI-enhanced neurotechnologies (McLennan et al., 2020). By integrating ethical considerations from the outset, the embedded ethics approach ensures that ethics is not just reactive to technological innovation, but proactively steers such innovation into a societally desirable direction. This anticipatory ethos ensures that the technologies we develop serve society without infringing upon our foundational values and human rights such as autonomy, privacy, accountability and distributive justice.

In this book, we have traversed a landscape where the fusion of brains and machines is no longer a distant speculation but a concrete reality. The ethical symbiosis of these entities requires continuous scrutiny and a commitment to adapt our moral compass to the evolving capabilities of AI. As we move forward on this journey, it is imperative to cultivate an ethical framework that is as dynamic and reflective as the technologies it seeks to govern, ensuring that the harmonization of brains and machines is accomplished with the utmost respect for the well-being of human beings and the integrity of conscious life, in all its forms.

To navigate this emergent ethical terrain, a roadmap for collaborative and ethically responsible research must be articulated. This roadmap should prioritize the establishment of interdisciplinary consortia that unify neuroscientists, ethicists, technologists, legal experts, and societal stakeholders. Such collaboration is essential to ensure that diverse perspectives inform the ethical frameworks that guide research and innovation. Moreover, the development of international guidelines and standards for ethical AI and neurotechnology research can provide a harmonized approach, accommodating cultural and societal diversity while upholding universal human rights (Ienca, 2021; Ligthart et al., 2023). Transparent public engagement must also be a cornerstone of this journey, fostering an informed dialogue about the benefits and risks of AI and neurotechnological advancements (Das et al., 2022). Continuous ethical education and training for researchers and practitioners across different disciplines will also be crucial, equipping them to anticipate ethical dilemmas and engage in reflexive practice. Lastly, ethical audits and accountability measures should be embedded within the research lifecycle, not only to assess compliance with established norms but also to adaptively refine those norms in response to novel findings and societal shifts. This proactive and inclusive approach will chart a course for research that not only advances knowledge but also aligns with our collective moral compass.

The creation of such a unified ethical framework calls for a collaborative synthesis of expertise. It requires a dialogue that not only bridges the gaps between these domains but also unites them in pursuit of a common goal: to ensure that the integration of AI and neuroscience progresses in a way that is beneficial and respectful of human dignity, agency, and the broader ecosystem of sentient life. This integrated framework should serve as a compass to guide us through the ethical labyrinth of this new cognitive era, ensuring that the advancement of technology is matched with an equally robust moral and ethical evolution.

References

Das, J., et al. (2022). Neuroscience is ready for neuroethics engagement. *Frontiers in Communication, 7,* 909964.

Ienca, M., & Ignatiadis, K. (2020). Artificial intelligence in clinical neuroscience: methodological and ethical challenges. *AJOB neuroscience, 11,* 77–87.

Ienca, M., et al. (2022). Towards a governance framework for brain data. *Neuroethics, 15,* 20.

Ienca, M. (2021). On neurorights. *Frontiers in Human Neuroscience, 15,* 701258.

Ligthart, S., et al. (2023). Minding rights: Mapping ethical and legal foundations of neurorights. *arXiv preprint arXiv:2302.06281.*

Martinez-Martin, N., Insel, T. R., Dagum, P., Greely, H. T., & Cho, M. K. (2018). Data mining for health: staking out the ethical territory of digital phenotyping. *NPJ digital medicine, 1*, 68.

McLennan, S., et al. (2020). An embedded ethics approach for AI development. *Nature Machine Intelligence, 2*, 488–490.

Salles, A., Evers, K., & Farisco, M. (2020). Anthropomorphism in AI. *AJOB neuroscience, 11*, 88–95.

Printed in the United States
by Baker & Taylor Publisher Services